Sebastian Schmidt

Numerical simulation of granular flow

Sebastian Schmidt

Numerical simulation of granular flow

An interdisciplinary, hydrodynamic approach for dense and dilute regimes

Südwestdeutscher Verlag für Hochschulschriften

Impressum/Imprint (nur für Deutschland/ only for Germany)
Bibliografische Information der Deutschen Nationalbibliothek: Die Deutsche Nationalbibliothek verzeichnet diese Publikation in der Deutschen Nationalbibliografie; detaillierte bibliografische Daten sind im Internet über http://dnb.d-nb.de abrufbar.
Alle in diesem Buch genannten Marken und Produktnamen unterliegen warenzeichen-, marken- oder patentrechtlichem Schutz bzw. sind Warenzeichen oder eingetragene Warenzeichen der jeweiligen Inhaber. Die Wiedergabe von Marken, Produktnamen, Gebrauchsnamen, Handelsnamen, Warenbezeichnungen u.s.w. in diesem Werk berechtigt auch ohne besondere Kennzeichnung nicht zu der Annahme, dass solche Namen im Sinne der Warenzeichen- und Markenschutzgesetzgebung als frei zu betrachten wären und daher von jedermann benutzt werden dürften.

Verlag: Südwestdeutscher Verlag für Hochschulschriften Aktiengesellschaft & Co. KG
Dudweiler Landstr. 99, 66123 Saarbrücken, Deutschland
Telefon +49 681 37 20 271-1, Telefax +49 681 37 20 271-0, Email: info@svh-verlag.de
Zugl.: Kaiserslautern, TU, Diss., 2009

Herstellung in Deutschland:
Schaltungsdienst Lange o.H.G., Berlin
Books on Demand GmbH, Norderstedt
Reha GmbH, Saarbrücken
Amazon Distribution GmbH, Leipzig
ISBN: 978-3-8381-1043-1

Imprint (only for USA, GB)
Bibliographic information published by the Deutsche Nationalbibliothek: The Deutsche Nationalbibliothek lists this publication in the Deutsche Nationalbibliografie; detailed bibliographic data are available in the Internet at http://dnb.d-nb.de.
Any brand names and product names mentioned in this book are subject to trademark, brand or patent protection and are trademarks or registered trademarks of their respective holders. The use of brand names, product names, common names, trade names, product descriptions etc. even without a particular marking in this works is in no way to be construed to mean that such names may be regarded as unrestricted in respect of trademark and brand protection legislation and could thus be used by anyone.

Publisher:
Südwestdeutscher Verlag für Hochschulschriften Aktiengesellschaft & Co. KG
Dudweiler Landstr. 99, 66123 Saarbrücken, Germany
Phone +49 681 37 20 271-1, Fax +49 681 37 20 271-0, Email: info@svh-verlag.de

Copyright © 2009 by the author and Südwestdeutscher Verlag für Hochschulschriften Aktiengesellschaft & Co. KG and licensors
All rights reserved. Saarbrücken 2009

Printed in the U.S.A.
Printed in the U.K. by (see last page)
ISBN: 978-3-8381-1043-1

Für meinen Opa

Herbert Schneider

Contents

Introduction   5

1 Models   13
    1.1 The basic hydrodynamic model of fluid flow 14
         1.1.1 The generalized Navier-Stokes Equations 15
         1.1.2 Boundary conditions 16
    1.2 Kinetic theory of granular gases 17
    1.3 Modeling the dilute and dense flow of grains 19
         1.3.1 Kinetic modeling 20
         1.3.2 An attempt to bridge kinetic and plastic models 24
         1.3.3 Implemented model 27

2 Algorithms   33
    2.1 Discussion on the type of numerical method 35
         2.1.1 Arguments for an implicit, pressure based splitting . . . 35
         2.1.2 The need for a nonlinear method 37
         2.1.3 The treatment of the temperature equation 38
    2.2 General space discretization 39
         2.2.1 Notation 39
         2.2.2 Finite Volume space discretization 42
    2.3 Introduction to fractional step methods 48
         2.3.1 The variants of fractional step methods 48
         2.3.2 Linear pressure correction algorithm 50
    2.4 A nonlinear pressure based algorithm 52
         2.4.1 Time discretization 52
         2.4.2 Spatial discretization of the split system 56
         2.4.3 Derivation of the nonlinear pressure equation 58
         2.4.4 The Newton method and its variants 60
         2.4.5 The nonlinear pressure algorithm 61
         2.4.6 Detailed discretization 63
    2.5 Initial and boundary conditions 71
         2.5.1 Approximation of boundary conditions 71
         2.5.2 Initial conditions for the granular flow model 74

3 Validation and numerical simulations   77
    3.1 The granular shear flow experiment 77
    3.2 Validation of the granular flow model 78
         3.2.1 Shear flow 79
         3.2.2 The angle of repose 82

		3.2.3 Sliding down a rough inclined plane 83

           3.2.3 Sliding down a rough inclined plane 83
           3.2.4 The stress tensor for granular flow 85
   3.3 Validation of the algorithm 87
        3.3.1 Newtonian flow 87
        3.3.2 Solutions for different grid resolutions 89
   3.4 Numerical investigations 91
        3.4.1 Compressibility regimes 91
        3.4.2 Mass conservation 92
        3.4.3 Bifurcating solutions 93
   3.5 Simulation of industrial processes 94
        3.5.1 Emptying of silos 95
        3.5.2 Compactification of granular media 97

4 Software   101
   4.1 Architecture and components 102
        4.1.1 The framework components 103
        4.1.2 The implementations 105
        4.1.3 Discussion of the modular approach 106
   4.2 A generalized approach to discretization 109
        4.2.1 Grid and volume data structure 110
        4.2.2 The discretization process 112
   4.3 Parallel linear algebra 115
        4.3.1 MPI data structures 115
        4.3.2 Assembly of the matrix 116
        4.3.3 The nonlinear case 119
        4.3.4 Discussion on the efficiency of the parallelization . . . 120
   4.4 Multiphase 121
        4.4.1 Multiphase through coupling terms 123
        4.4.2 Multiphase through a coupled system 123

5 Concluding remarks and outlook   125

A Appendix   133
   A.1 Collaps in kinetic models 133
   A.2 Dynamic Coulomb friction 134

Notation   137

Bibliography   148

Introduction

Granular materials show a wonderfully diverse set of behaviors. Make a sand castle, and the material appears solid. Push on the castle and it can fall down in an avalanche-like pattern. Sometimes the avalanche moves the bulk of the material, sometimes it is confined to a thin layer on the surface. Shake up crushed ice in a martini shaker, and it moves like a gas. Try to pour salt through an orifice, and it has a characteristic tendency to choke up and clog the orifice. Gas, liquid, solid, plastic flow, glassy behavior - a granular material can mimic them all. In addition, the properties of a granular material can depend upon its history. Tamped sand is different from loose sand. But in many ways, a granular material is like an ordinary fluid. Both types of material are composed of many small particles, and each has a bulk behavior that hides the materials graininess. It is thus natural to ask whether the same equations, concepts, and theories that work for molecular material also apply to the granular form of matter.

LEO P. KADANOFF, "BUILT UPON SAND"[1]

About this work

The goal of this work is the simulation of granular flow. The definition of "granular flow" is a nontrivial task in itself, see for example [Dar03]. We say that it is either the flow of grains in a vacuum or in a fluid. A grain is an observable piece of a certain material, for example stone when we mean the flow of sand.

Choosing a hydrodynamic view on granular flow, we treat the granular material as a fluid. A hydrodynamic model has to be developed, that describes the process of flowing granular material. This is done through systems of partial differential equations (PDEs) and algebraic relations. Solutions to these systems have to be obtained to understand the process. The equations are

---

[1][Kad99]

in most cases so difficult to solve that an analytical solution is out of reach. So approximate solutions must be obtained.

Hence the next step is the choice or development of a numerical algorithm to obtain approximate solutions of the model. Common to every problem in numerical simulation, these two steps do not lead to a result without implementation of the algorithm. Hence the author attempts to present this work in the following frame, to participate in and contribute to the three areas Physics, Mathematics and Software implementation and approach the simulation of granular flow in a combined and interdisciplinary way.

This work is structured as follows. A continuum model for granular flow which covers the regime of fast dilute flow as well as slow dense flow up to vanishing velocity is presented in Chapter 1. This model is strongly nonlinear in the dependence of viscosity and other coefficients on the hydrodynamic variables and it is singular because some coefficients diverge towards the maximum packing fraction of grains. Hence the second difficulty, the challenging task of numerically obtaining approximate solutions for this model is faced in Chapter 2. In Chapter 3 we attempt to validate both the model and the numerical algorithm through numerical experiments and investigations and show their application to industrial problems. We finish with the implementation of the simulation tools we have developed in Chapter 4.

Throughout the whole work, we focus on one experiment as our guideline. This is the shear flow experiment from [BLS+01]. It serves well to demonstrate the algorithm, all boundary conditions involved and provides a setting for analytical studies in [BLS+01] to compare our results.

Review and state of the art

Modeling of granular flow: We attempt to model the flow of granular material as a liquid. Due to the lack of a clear time and spatial scale separation and very efficient mechanisms for energy dissipation in form of inelastic collisions (see [Kad99]) this may seem overly brave. For mainly these reasons, a theory is still missing for the description of granular flow on a macroscopic scale with the same accuracy as the Navier-Stokes Equations (NSE) for simple liquids. Still kinetic theory and hydrodynamic modeling are in many cases a valid approach to the simulation of granular flow, see [Duf01].

We have to trust the usual assumption that even for granular materials, the spatial variation of the hydrodynamic variables can be captured by terms linear in spatial derivatives. Then fortunately, for weakly inelastic granular media, kinetic theory provides a framework for deriving the correct hydrodynamic equations presented in [BP03]. The kinetic theory with heuristic modifications is very useful for many simulations of granular flow in application problems at intermediate volume fractions as for example in the simulation of fluidized beds in [Gid94].

Kinetic theory assumes that grain collisions are binary which means that collisions always occur only between two particles at the same time and instantaneous which means that the particles separate immediately after the collision, see [BP03, Section 1.4, pg. 5]. It seems obvious at first sight that this assumption is invalid for volume fractions close to the maximum packing. Therefore, the applicability of kinetic theory becomes questionable in this regime. Nevertheless, the literature reports differently. In [MPB03], Meerson et al. show strikingly good agreement between simulations at large volume fractions and hydrodynamic theory. The same has been done by Boqcuet et al. in [BLS$^+$01, BEL02] relying on experiments. These experiments were carried out under shearing conditions which will become important later in this work. In that case they have shown that kinetic theory is able to mimic solid like behavior. It does so by exhibiting a solid like Coulomb stress as a solution of the hydrodynamic equations.

Numerics: We model granular flow by a time-dependent NSE-type system. To compute approximate solutions of the system we need to discretize it in space and time. The former is achieved using a Finite Volume (FV) discretization. The latter is a novel pressure based nonlinear fractional step method (NFSM). In the linear case, we call every method which decouples the solution of the NSE-type system into multiple steps a linear fractional step method (LFSM). This may be misleading from a steady flow point of view because fractional step methods usually split the time step. However, we are mainly concerned with the unsteady case where all these methods can be interpreted as splitting one full solution step into fractional steps.

The FV discretization seems to have been first introduced by Tikhonov and Samarskii in [TS61]. It partitions the domain into volumes over which the equations are integrated. The resulting volume integrals are transformed to surface integrals along the surface of the volume where possible. The approximation of these surface integrals then yields the FV discretization. A collection of recent FV discretizations and the state of the art mathematical theory of FV discretization schemes can be found in [EGH00].

The main concern of this work will be the time discretization of the NSE-type granular flow model. Considering only implicit or semi-implicit methods which decouple the NSE-type system, the first developments that lead to pressure based LFSMs go back to the works of Chorin in [Cho68] and Patankar and Spalding in [PS72]. Chorin has proposed a projection method for unsteady incompressible flow problems which is basically a two-stage LFSM. Patankar and Spalding have introduced a FV implementation of these LFSMs in the SIMPLE method for steady incompressible flow. The underlying idea of all the methods is the projection of a predicted velocity and pressure onto the space of divergence free velocity.

We see that pressure based methods were initially developed for incompressible flow. The first extension of pressure based schemes for weakly compressible flow dates back to [HA68, HA71]. Much further work has been published on pressure based LFSMs for the unsteady, compressible, non-isothermal NSE which are the basis for our equation system. For an overview of methods see the book of Ferziger and Perić [FP96, Chapter 7] and for examples of recent methods see [vVW03] or [Chu03].

Mathematical frameworks have been used to systematically describe pressure based methods. The Schur-Complement (SC) notation discussed by Turek in [Tur99] and operator splitting discussed in [GS98] should be mentioned here. These attempt to provide a way to compare the many variants of pressure based LFSMs.

Motivation, goals and overview

It becomes clear that the simulation of granular flow inherits two distinct difficulties - the modeling and the numerical treatment. Regarding the first, there is still no common agreement on the mathematical description for this type of flow. Regarding the second, numerical methods that are able to solve the presented model in both the dilute and the dense regime are rare.

Modeling of granular flow: We have stated above that in certain situations current hydrodynamic models from kinetic theory show very good agreement with experiments. Let us try to explain the reasons to be able to understand where problems will occur in other situations.

The mentioned experiments and simulations are carried out under shearing conditions. This is a situation of permanent input of energy through the moving wall that shears the material. Under these conditions, the existence of a dynamic Coulomb stress might explain why molecular dynamics (MD) simulations and experiments can be reproduced by hydrodynamic theory at all. For flowing granular media close to maximum packing fraction, collisional contacts of grains are replaced by frictional contacts. The proper theory of the stress in that case would be some theory of static friction, for example the Coulomb friction theory. However, when permanent energy input prevents the granular system from arresting, the dynamic Coulomb friction of the hydrodynamic theory is able to mimic the true Coulomb friction.

The nature of these experiments hides a flaw in the kinetic models. This flaw surfaces only when a force is missing that would prevent the system from arresting. We will show in Section 1.3.2 that the above arguments will become invalid in that case. This motivates the extension of the available models.

We will show that kinetic theory alone is not able to reproduce the qualitative behavior of typical arresting processes as for example the formation of heaps. In Section 1.3 we develop a model that is able to describe also arresting granular flow. Motivated by the work of Savage [Sav98], we present a hybrid model of kinetic theory and a theory derived from soil mechanics. This theory overcomes the difficulties observed in kinetic theories and extends the applicability of hydrodynamic theory to arresting granular flow.

Our model is a quite simplified version of the one presented by Savage. There is a good reason for that. The wealth of constitutive models is astonishing, see [Kol00] and a correct model has not been identified. It is necessary to obtain a constitutive model which can be calibrated and has as few parameters as possible. We present such a model and show that we can produce the same results using our simplified model. We test the theory that we introduce by simulating heap formation with predictable angle of repose, comparing simulations of flow down an inclined plane and reproducing core and mass flow in silos.

Numerics: The presented model is a strongly nonlinear and singular system of PDEs and constitutive relations. To numerically obtain approximate solutions of this system we discretize it in space by using the FV method on a cell-centered grid of cuboids with collocated arrangement of the unknowns. The discretization is derived from the integrated equations. Convective terms are discretized by a first order upwind scheme and second order derivatives are discretized using central differences.

The main focus of this work lies in the discretization of the system in time by a pressure based NFSM. We start with a discussion on this numerical approach. We will especially point out our reasons for an implicit pressure based algorithm and discuss that the nature of the model strongly suggests the use of a nonlinear method. We then introduce the general notation and the operators for the space discretization of certain PDE terms that we will use throughout this work. We proceed with an introduction to LFSMs and the derivation of a linear pressure correction algorithm (LPCA) which will form the basis of the derivation of our NFSM.

Based on this derivation we introduce the splitting of the coupled granular flow system in a few easier to solve problems and derive a nonlinear pressure equation (NPE) from the split system. The first step will be the prediction of a velocity field from a linearized, fully implicit momentum equation using an old pressure field. Then we will present the derivation of a coupled system of a nonlinear pressure equation and a velocity correction equation as the correction step.

Finally we describe the nonlinear pressure algorithm (NPA) for solving the full time-dependent system. For the solution of the NPE we use a truncated Newton method for systems of nonlinear equations. This method requires the evaluation of the Jacobian of the NPE. We obtain the Jacobian by finding the derivatives of the NPE and give a detailed discretization of both the NPE and the Jacobian.

Chapter 1

Models

This chapter provides all models and modeling aspects of the thesis. The models are derived on the basis of a hydrodynamic description of flows and are of Navier-Stokes Equations (NSE)-type. For granular flow, this assumes that a continuum description of particles is allowed which is discussed in [BP03, Section 1.3]. They argue that particles of granular gases are macroscopic bodies which allows for a continuum description of their interactions through a stress-strain relation.

Our aim is to arrive at a set of equations which models the flow of grains in both dilute and dense regimes. The former is often called the regime of rapid granular flow and is extensively discussed in the literature, see [Duf01]. It seems to be agreed that for this regime hydrodynamic models are able to reproduce many phenomena of granular flow and the constitutive relations based on kinetic theory are valid. Outside this regime the topic becomes controversial. Clearly, a hydrodynamic model will not be able to reproduce mechanical interactions of single grains. However, we do argue that hydrodynamic equations for granular flow are able to model the flow of dense bulk material as long as there is at least some movement.

The idea presented to arrive at such a model is a crossover of the constitutive relations locally depending on the flow regime which depends on the volume fraction. The differential equations are the same throughout the regimes. Only the relations for viscosity, pressure, granular temperature etc. are continuously changed for the dense regime. This introduces a certain arbitrariness because it is not clear where dilute flow ends and where dense flow begins. It must be asked at what volume fraction the flow is not dominated anymore by instantaneous collisions of grains but mainly by the sliding and rolling of grains on each other. We can not answer this question, we have to treat the crossover volume concentration as a parameter which must be identified by comparison with experiments.

In Section 1.1 we shortly mention the commonly known system of generalized NSE for the description of compressible non-isothermal fluid flow with varying viscosity. Then we proceed to Section 1.2 where we will provide a very brief introduction to the kinetic theory of granular gases. This is followed by the derivation of a model for granular flow based on that theory in Section 1.3. Our final model extends the description of granular gases of [BP03]. In the regime above a certain volume fraction of grains where the assumptions of kinetic theory are not valid, we use a modeling approach from soil mechanics similar to the work of [Sav98].

Let $1 \leq d \leq 3$ be the space dimension. Then as usual $\mathbf{x} \in \mathbb{R}^d$ and $t \in \mathbb{R}$ denote the space and time coordinates respectively. Let us denote velocity by $\mathbf{u} = \mathbf{u}(\mathbf{x}, t)$, pressure by $p = p(\mathbf{x}, t)$, density by $\rho = \rho(\mathbf{x}, t)$ and volume forces by $\mathbf{f} = \mathbf{f}(\mathbf{x}, t)$. We work in arbitrary domains $\Omega \subset \mathbb{R}^d$ bounded by solid walls as well as inflow and outflow boundaries denoted by $\partial\Omega_{\text{sw}}$, $\partial\Omega_{\text{in}}$ and $\partial\Omega_{\text{out}}$ respectively.

1.1 The basic hydrodynamic model of fluid flow

In the later sections of this chapter we will present a model for granular flow which is based on the modeling of fluids. Therefore we give a short overview of the basic equations of fluid flow, the NSE. For detailed descriptions see [FP96] or [Wes01].

The NSE describe the motion of substances that can flow. This is based on several assumptions made on the fluid. The first is that the fluid is continuous. It signifies that it does not contain voids formed, for example, by bubbles of dissolved gases. Also this means that it does not contain aggregates of any sort of particles. Another necessary assumption is that all the scalar and vector fields like velocity, pressure, density and temperature are differentiable. This would not be the case for, say, phase transitions.

Given that all these assumptions are valid for the fluids we consider, the equations are derived from the basic principles of conservation of mass, momentum and energy. Usually this is done by considering a finite arbitrary volume called a *control volume* over which these principles can be easily

1.1. THE BASIC HYDRODYNAMIC MODEL OF FLUID FLOW

applied. This control volume is fixed in time and space with flow allowed to occur across the boundaries. The NSE then follow from the conservation laws and linear constitutive relations. For a detailed derivation of this form see [Fle91a, Sections 11.2.1-11.2.4].

1.1.1 The generalized Navier-Stokes Equations

Let us start with the model for instationary, compressible, viscous, isothermal flow described by the system of NSE (1.1) for the unknowns $\mathbf{u}$, p and ρ.

$$\partial_t(\rho) + \operatorname{div}(\rho \mathbf{u}) = 0, \tag{1.1a}$$

$$\partial_t(\rho \mathbf{u}) + \operatorname{div}(\rho \mathbf{u} \otimes \mathbf{u}) - \operatorname{div}(\sigma) + \operatorname{grad}(p) - \mathbf{f} = 0, \tag{1.1b}$$

where

$$\sigma_{ij} := 2\eta\kappa_{ij} - \frac{2}{3}\eta\delta_{ij}\operatorname{div}(\mathbf{u}) \quad \text{with} \quad \kappa := \frac{1}{2}\left(\frac{\partial u_i}{\partial x_j} + \frac{\partial u_j}{\partial x_i}\right). \tag{1.1c}$$

Assuming that we can write the momentum as the product of density and velocity, Equations (1.1a) and (1.1b) are the conservation of mass and momentum respectively. Equation (1.1c) denotes the general stress tensor and the general, symmetrized rate of strain tensor. Volume forces are given in the right hand side $\mathbf{f}$ of Equation (1.1b). Furthermore, by δ we denote the Kronecker symbol. Though unusual for the standard NSE, we allow the viscosity to vary such that $\eta = \eta(\mathbf{x}, t)$ is our dynamic viscosity.

Due to the compressibility, the dependence of ρ on p, System (1.1) has to be extended by a relation between ρ and p which, for an ideal gas, reads

$$p = \rho R \hat{T}. \tag{1.2}$$

Here $\hat{T}$ is a given constant temperature and R a constant dependent on the fluid.

For non-isothermal flow, where T is not constant, a partial differential equation (PDE) is necessary to describe the dynamic behavior of the internal energy of the system. We do not consider this topic further at this point as it is treated extensively for Newtonian flow, for example in [Fle91a, Section

11.2.4]. It becomes an essential part of modeling for granular flow which we will discuss in the following section.

We have stated the NSE in their full complexity because we will need them as the basis of our model for granular flow. However, for validation purposes and for reference in later parts of this work, let us retreat to a more simple case of fluid flow. If we assume the flow to be incompressible with constant density $\hat{\rho}$, stationary and with constant dynamic viscosity $\hat{\eta}$ then System (1.1) is reduced to

$$\operatorname{div}(\hat{\rho}\mathbf{u}) = 0, \tag{1.3a}$$

$$\operatorname{div}(\hat{\rho}\mathbf{u} \otimes \mathbf{u}) - \operatorname{div}(\tilde{\sigma}) + \operatorname{grad}(p) - \mathbf{f} = 0, \tag{1.3b}$$

where $\mathbf{u}$ and p are sought, ρ and $\mathbf{f}$ are given and $\tilde{\sigma}$ is a much more simplified version of the stress tensor

$$\tilde{\sigma} = \hat{\eta}\tilde{\kappa}, \quad \tilde{\kappa}_{ij} := \frac{\partial u_i}{\partial x_j}. \tag{1.3c}$$

This models for example the stationary flow of water. Between System (1.1) and System (1.3) which will both be referred to in this work lie many orders of complexity. One of them is the model for instationary compressible Newtonian flow where the density is allowed to vary with pressure and the thermodynamic variables velocity, density and pressure are allowed to vary in time. We will develop our algorithm to solve the most complex model based on ideas for that last mentioned model.

1.1.2 Boundary conditions

Let us remark on the complicated issue of boundary conditions on open boundaries (inflow, outflow) for NSE. Because the NSE are of mixed type they have properties of the parabolic variant of the Stokes system for creeping incompressible flow, see [Wes01, p. 32, Equations (1.15),(1.36)] and the Euler equations for compressible convective flow, see [Wes01, p. 32, Equations (1.76)-(1.78)]. Boundary conditions for the Stokes system are treated well and most extensively in [EG04, Chapter 4]. Boundary conditions for the hyperbolic Euler equations are analyzed using the method of characteristics, see [Wes01, Section 10.2, p. 402ff].

The study of well-posed boundary conditions for NSE-type systems relies initially on the work of Strikwerda in [Str76] on boundary conditions for the class of incompletely parabolic equations which the NSE belong to. This was followed by [GS78]. In both works, the system of NSE is linearized and boundary conditions are chosen such that the energy of the system which is the square of the solution in characteristic variables decays exponentially in time on the boundary. Then well-posedness can be proven, see [OS78]. A different, completely nonlinear approach that does not rely on the framework of incompletely parabolic problems is followed in [Dut88]. All these works provide theoretical boundary conditions, but implementation is not straightforward from any of them.

Two works that give numbers of boundary conditions and forms that are implementable in a direct way are [HG97] and [SK03]. They state that for the 3-dimensional NSE (including an equation for internal energy) with subsonic inflow one must provide 5 boundary conditions on the inflow and 4 on the outflow. See [SK03, Table 1, Table 2]. One possible exact form of boundary conditions, though in characteristic variables is given in [HG97, Equations (19),(21)]. Based on these insights, we will provide in Section 1.3.3 the specific boundary conditions for the granular flow model.

1.2 Kinetic theory of granular gases

Our approach to modeling granular flow in the upcoming Section 1.3 is largely based on the kinetic theory of granular gases discussed in [BP03]. This introduction will be based on that book.

Classic kinetic gas theory is derived on the basis of molecule collisions. It looks at the microscopic properties of molecule interactions and from that derives a macroscopic description. One of the main assumptions is that the collisions are elastic which means that the relative velocities of molecules before and after collision have the same magnitude.

Because the collisions of particles or grains are inelastic, this basic assumption is not valid and the concepts of kinetic gas theory have to be generalized to account for the hence dissipative collisions. With these generalizations, granular gases may also be described by time-dependent fields of macroscopic variables such as pressure, density and temperature. For this one has to first investigate pairwise particle collisions. Then statistical properties of ensembles of such particles have to be derived in ways very similar to molecular gases.

In [BP03, Chapter 1] the mechanics of particle collisions are investigated. The starting point is the investigation of the collision of particles on a line where a coefficient of restitution e smaller than 1 quantifies the change in relative velocities before and after a collision. If two those particles have a relative velocity u_{12} before collision, the relative velocity after collision u'_{12} is given by

$$u'_{12} = eu_{12}, \quad \text{with} \quad e < 1. \tag{1.4}$$

The coefficient of restitution (which in [BP03] is called ε) serves as the central characteristics of a granular gas.

Through generalizations of the coefficient of restitution for collisions in space (see Figure 1.1, left) and further through few-particle systems, it is used to describe the cooling of granular gases due to dissipative collisions in [BP03, Chapter 5]. The macroscopic *granular temperature* is introduced as the average kinetic energy of a resting system of grains

$$T := \frac{1}{3} < \mathbf{u}^2 > \tag{1.5}$$

just as the temperature in the theory of molecular gases. For moving granular systems the granular temperature is the mean square velocity fluctuation of grains. From the energy loss of collisions in a collision cylinder as in the right side of Figure 1.1 the evolution of granular temperature is derived. Finally, for the case of constant coefficient of restitution it is shown in [BP03, Equation (8.12)] that the granular temperature behaves as

$$\frac{\partial T}{\partial t} \propto -C(\rho)\sqrt{T} \cdot T. \tag{1.6}$$

Above displayed is a simplified version of the relation presented in [BP03].

1.3. MODELING THE DILUTE AND DENSE FLOW OF GRAINS

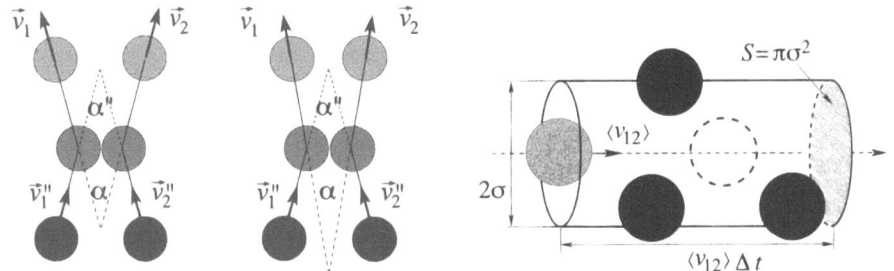

Figure 1.1: Left: [BP03, Figure 6.2] showing the difference in collision behavior in space and time for elastic (left) and inelastic (right) collisions. Right: [BP03, Figure 5.1] showing the collision cylinder. Only particles located within the collisions cylinder collide with the gray particle. From this the average energy loss through dissipative collisions is derived.

The important note is that the cooling of granular temperature depends only on the temperature and the density of the ensemble of grains. That latter dependency is collected in the function C.

Following these derivations, the hydrodynamic equations for granular gases are derived from the Boltzmann Equation and stated in [BP03, Section 17.3]. Furthermore, in the work of of [BLS$^+$01] a much simpler form of the equations is validated for a shearing experiment. These two works form the basis for large parts of the model we will derive in the following Section.

1.3 Modeling the dilute and dense flow of grains

We introduce a unified, hybrid, hydrodynamic model for granular flow. By unified we mean that the model aims to be valid in regimes of both dilute and dense granular flow. It will be hybrid because it will in parts interpolate existing modeling approaches between regimes. It will be a hydrodynamic approach because our modeling will results in a system of the type of System (1.1) assuming that we can describe the ensemble of particles as a continuum. We account for the granular aspects with a regime dependent constitutive theory for inelastic particles.

A study of available results in the literature shows the manifested separation of the modeling of granular flow. Dilute granular flow is often modeled using kinetic theory as in [BP03]. Dense granular flow, if modeled at all on the basis of the NSE, is usually modeled with quasi-static approaches as in [GD99]. We model granular flow in both regimes. We keep the same differential equations through all regimes and vary the constitutive relations locally and continuously depending on whether we are in the dilute or dense regime.

Our constitutive model will be non-Newtonian with the following characteristics. First, the dynamic viscosity η depends on time and space and will increase largely with increasing volume fraction. Secondly we introduce an equation for granular temperature T which will model the process of heating and cooling of the granular material as in Section 1.2. And we introduce a relation between granular pressure p, granular density ρ and T.

We start with relations from the kinetic theory of granular gases in Section 1.3.1. We will show that these are not sufficient to describe arresting granular flow correctly in the high density regime. Based on these insights, we will extend the kinetic model by special relations for the dense regime in Section 1.3.2.

1.3.1 Kinetic modeling

A granular material differs from a simple fluid in many aspects. The most obvious is that energy is not conserved on the scale of grains. Contrary to gas dynamics, collisions between grains are inelastic, see Section 1.2 and (1.4). Parts of the kinetic energy of the grains before the collision is transferred to the molecules making up the grain, hence slowing down the colliding grains. To capture this effect, we have defined a granular temperature T in (1.5) which measures the fluctuating random motion of the grains in (1.5) as in [BP03, Section 5.1]. We would like to stress again that the dissipation of granular temperature is a crucial feature to the kinetic modeling of granular flow. It models the difference between molecular and granular gases. This is, a closed system containing a molecular gas remains in its initial state of average motion where a granular gas tends towards the state of resting grains with zero granular temperature.

The equation for granular temperature

The equation for the granular temperature is derived from the Boltzmann equation in the ensemble of constant volume in [BP03, p. 52ff]. Using standard techniques from [LL78], the resulting equation needs to be transformed into the ensemble for constant pressure, as this is valid in our situation, see [Lat06]. As mentioned before, the dynamics of the granular temperature can be derived from the Boltzmann Equations, see [BP03, Chapter 5 and Chapter 17, Equation (17.32)].

$$\rho \partial_t(T) + \rho \mathbf{u} \operatorname{grad}(T) = \frac{2}{3}(\sigma : \kappa - \operatorname{div} \mathbf{q}) - \underbrace{\rho \varepsilon T}_{T_{\text{diss}}}, \qquad (1.7)$$

with the heat flux $\mathbf{q}$ and the energy loss rate ε.

Equation (1.7) has the usual form of a heat transport equation as in [LL78] with T_{diss} added. The term T_{diss} describes the dissipation of granular temperature due to inelastic collisions on the basis of Equation (1.6). The left hand side describes the change of granular temperature due to free streaming. The right hand side describes the effects of diffusive temperature transport, viscous heating and dissipation respectively.

Constitutive equations from kinetic theory

System (1.1a),(1.1b),(1.7) consists of three equations with the unknowns $\rho, \mathbf{u}, p, T$ and $\sigma, \mathbf{q}, \varepsilon$. We obtain a closed model by providing constitutive relations for these variables. Small and intermediate densities are well described by the hydrodynamic equations derived by kinetic theory. Therefore, the derivation of hydrodynamic equations as well as expressions for the stress tensor σ and all transport coefficients is possible. In [BP03] and [GD99], the hydrodynamic equations are derived using Chapman-Enskog theory. Hence, the stress tensor and the heat flux are in lowest order linear in gradients of the hydrodynamic variables

$$\tilde{\sigma} = \eta \tilde{\kappa}, \quad \mathbf{q} = -(\lambda \operatorname{grad}(T)). \qquad (1.8)$$

Here λ denotes the heat conductivity and $\tilde{\kappa}$ denotes the non-symmetrized strain rate tensor from (1.3c) given by

$$\tilde{\kappa}_{ij} = \frac{\partial u_i}{\partial x_j}. \tag{1.9}$$

Let us explain the use of a non-symmetric stress tensor $\tilde{\sigma}$. A non-symmetric stress tensor violates the conservation of angular momentum of the macroscopic flow field. This is very hard to justify for simple liquids as the molecules making up the liquid do not have additional macroscopic rotational degrees of freedom. For granular media however the situation is different. Macroscopic sources of rotation are caused by the microscopic dissipation of energy (1.4). This violation of macroscopic angular momentum found in [Dah59], [Cam93] is used in the modeling of collisional granular flow in [MHN02]. The change of tangential velocity by grain collisions is discussed in [BP03, Section 3.4]. There it is shown that even for ideally smooth but inelastic spheres the coefficient of tangential restitution is less than 1. This means that in an irreversible process, the violation of the conservation of energy, caused by inelastic collisions, energy is dissipated into sources of macroscopic rotation.

In this context, we derive the non-symmetrized stress tensor by defining a rotational viscosity η_R equal to the shear viscosity η. The definition of the stress tensor is then given as

$$\sigma = \frac{\eta}{2}\left(\tilde{\kappa} + (\tilde{\kappa})^T\right) + \frac{\eta_R}{2}\left(\tilde{\kappa} - (\tilde{\kappa})^T\right),$$

which equals $\tilde{\sigma}$. Finally we want to point out that for the incompressible case, assuming constant viscosity, the part $(\tilde{\kappa})^T$ of the strain rate tensor does not contribute to the accelerating force at all. This is due to

$$\frac{\partial \eta(\tilde{\kappa})_{ij}^T}{\partial x_j} = \eta\frac{\partial u_j}{\partial x_i \partial x_j} = \eta\frac{\partial \operatorname{div} \mathbf{u}}{\partial x_j} = 0. \tag{1.10}$$

In the dense regime we only have weakly compressible flow. In that case, Equation (1.10) suggests that our approximation of the stress tensor does not influence the results too significantly. This form of the stress tensor can also be justified with numerical experiments, see Section 3.2.4.

1.3. MODELING THE DILUTE AND DENSE FLOW OF GRAINS

The expressions for the transport coefficients also follow from kinetic theory. They are derived in [GD99] and their form is quite involved. Fortunately Bocquet and others show in [BLS+01] and [BEL02] that a much simpler form of the equations can produce quantitatively correct results for a shearing experiment. The necessary criteria, to preserve the low and high density limits is fulfilled even for this simple form. We will furthermore show in Chapter 3 that this more simple form together with our extensions is applicable to various regimes of granular flow.

Hence we choose a similar form as in [BLS+01] for the transport coefficients and the constitutive relation for pressure. We denote these expressions by the subscript K because they are derived from kinetic theory. They read

$$p_K = T g(\rho) \rho \tag{1.11a}$$
$$\varepsilon_K := \varepsilon_0 \rho^2 \sqrt{T} g(\rho) \tag{1.11b}$$
$$\lambda_K := \lambda_0 \sqrt{T} g(\rho) \tag{1.11c}$$
$$\eta_K := \eta_0 \sqrt{T} g(\rho), \tag{1.11d}$$

where for the maximum packing density ρ_c,

$$g(\rho) := \left(1 - \frac{\rho}{\rho_c}\right)^{-1}. \tag{1.12}$$

Here g is the value of the radial distribution function at contact for a given density. It models the tendency of grains to touch between low and high density and diverges at ρ_c. The constants ε_0, λ_0 and η_0 are specified in Section 1.3.3. As we will see later, the introduced model is able to reproduce effects that make granular materials quite different from fluids. One this the ability to form piles. This is due to a property of the model we call dynamic Coulomb friction.

Dynamic friction angle

One very interesting property of the model is that an internal friction angle is implicitly contained in th solution of the model. It is shown in [BLS+01,

Equation (35)] that for the high density limit, which is the only case where we would expect the formation of piles

$$\sigma_{x_1 x_2} \propto \mu_0 p$$

where $\sigma_{x_1 x_2}$ is the shear stress in x_2-direction when shearing in x_1-direction and μ_0 is some factor which is given in detailed form in [BLS+01, Equation (35)]. This means that the shear force is expected to only depend on the pressure and in particular to be independent of the shear rate κ as is usually found in solid friction.

This model is used successfully for rapid granular flow and also for intermediate densities. However, in the high density limit its validity becomes questionable. We will consider this in the following Section.

1.3.2 An attempt to bridge kinetic and plastic models

By the definition of the granular temperature from kinetic theory in (1.5), a resting bulk of grains has zero granular temperature. However, a bulk of grains rests because the gravity force is balanced by the repulsive part of the potential energy of the molecules constituting the grains. We will show in the next section that this is missing in the kinetic model. That is the reasons why kinetic theory is not able to reproduce this static situation in a stable way. Consequently we need to simulate a potential energy. That will be done by introducing an athermal pressure together with a constant nonzero granular temperature for bulk material based on critical state theory from soil mechanics.

Shortcomings of the kinetic model

The kinetic model is usually validated and produces good results in situations like shear flow and gravity driven flow as in [BLS+01] and [DD99]. In both these situations it is ensured that energy is continuosly put into the system. This happens either via the application of a permanent torque in the shearing experiment or the permanent transformation of potential energy into kinetic energy of the flow in the gravity driven flow. This guarantees that the granular temperature always stays non-zero which in turn allows the purely thermal kinetic pressure p_K from Equation (1.11a) to stabilize the system.

However, kinetic theory does not account for the strongly repulsive forces between the grains, except for the radial distribution function at contact g from (1.12), which is the Enskog correction to pressure and transport coefficients. It models the repulsion of grains at a certain density. If for any reason, the temperature reaches the zero limit faster than the density reaches the maximum packing limit then the pressure tends to zero and there is no force preventing the system from collapsing. This phenomenon called inelastic collapse is considered in [BP03, Section 4.1] for the more simple case of a chain of particles. It is shown there that initial conditions for a chain of particles exist for which the energy of the relative motion is completely dissipated. In Appendix A.1 we will investigate this using Haff's homogeneous cooling.

In addition to this instability, there are also indications in [DMB+03], that for high densities and small granular temperatures, the viscosity should not decrease with decreasing temperature as in Equation (1.11). It should rather dramatically increase. This discrepancy is due to neglecting collective phenomena caused by the strong repulsion of the grains.

Hybrid constitutive model

The origin of the problems described in the previous section is the lack of any static force in the kinetic constitutive theory. Such a force would reflect the hard core repulsion of the grains and is hence necessary. Macroscopically this force is felt as the impossibility to compress a resting granular medium beyond a limit of density or a resistance against external pressure. The dynamics in those situations is mostly associated with plastic deformations as described by soil mechanics. A first attempt to a model bridging kinetic and plastic regimes dates back to Savage [Sav98]. We will adopt a simplified model of [Sav98] which nevertheless captures the essential features and is capable of reproducing many known experimental results of granular flow from the dilute to the dense regime.

The main idea is to introduce a contribution p_Y to the pressure which is independent of the granular temperature. This contribution is acquired only above a certain cross over density ρ_{co}. In soil mechanics, one introduces a yield surface above which quasi-static deformations occur. The pressure p_Y is related to the pressure on the yield surface as in [Sav98]. A discussion of the literature on the functional dependency of the yield pressure on the density can also be found in [Sav98]. The exact form is not known. Hence, for simplicity, it is acceptable to choose the same form as for the kinetic pressure. The yield pressure p_Y is given by

$$p_Y = \Theta(\rho - \rho_{co}) \cdot T_0 \cdot (\rho - \rho_{co}) g(\rho), \tag{1.13}$$

where Θ is the Heaviside step function. The constant T_0 provides the non-vanishing athermal pressure which assures stability in the quasi-static regime. Then the total pressure is the sum of both pressures

$$p = p_K + p_Y. \tag{1.14}$$

The transport coefficients η, ε and λ also have to be modified. They need to fulfill the following requirements:

1. They must not vanish with vanishing temperature.

2. The viscosity η has to diverge with vanishing temperature. This is consistent with glass transition which dense granular media is observed to resemble at $T = 0$, see [BLS+01].

3. The crossover from the kinetic regime to the yield regime must not modify the internal friction angle. Otherwise stable piles would start to become instable when regimes are crossed.

We fulfill these requirements with the following relations for viscosity, energy

1.3. MODELING THE DILUTE AND DENSE FLOW OF GRAINS

dissipation rate and heat conductivity

$$\eta := \eta_K(1 + \frac{p_Y}{p_K}), \tag{1.15a}$$

$$\varepsilon := \varepsilon_K(1 + \frac{p_Y}{p_K}), \tag{1.15b}$$

$$\lambda := \lambda_K(1 + \frac{p_Y}{p_K}). \tag{1.15c}$$

Point 1 is clearly fulfilled because by the definition of p_Y (1.13) the yield pressure is independent of the temperature. For the third point, see Appendix A.2 for details on the dynamic Coulomb friction and the friction angle for our hybrid model.

The analysis of the second point is a much more involved issue. For the case of homogeneous cooling we show in Appendix A.1 that for the purely kinetic relations (1.11) the temperature tends towards zero faster then the density reaches the limit ρ_c and hence the kinetic pressure p_K tends towards zero. In this case requirement 2 above is fulfilled because p_Y (1.13) is finite.

However, it is not clear if the homogeneous cooling is always the case. As the decay of temperature and the limit behavior of density is coupled within the full model and in particular the equation of granular temperature (1.7), the full analysis of this issue is out of the scope of this work.

This closes the hybrid model for dilute and dense granular flow. The specific details that remain untouched will be discussed in Section 1.3.3 where we list the complete closed model together with implementational details.

1.3.3 Implemented model

The temperature Equation (1.7) from [BP03, Equation (17.32)] is derived in non-conservative form. This is consistent because the granular temperature is not a conserved quantity. However, for a Finite Volume (FV) discretization we want an equation for the unknown ρT rather than just T. Hence we use Equation (1.1a) to change the term involving the time derivative and the convective term such that

$$\rho \partial_t(T) + \rho \mathbf{u}\,\text{grad}(T) \stackrel{(1.1a)}{=} \partial_t(\rho T) + \text{div}(\rho \mathbf{u} T).$$

We would like to comment on the issue of unphysical granular heating due to the dissipative term that we have found for our implemented model in numerical experiments. Let us consider the case of very dilute granular flow, for example when a domain is initially supposed to be empty of granular material which we simulate by providing a very low initial volume fraction, compare for Section 2.5.2.

Kinetic theory in [BP03] yields that the dissipative term T_{diss} in Equation (1.7) depends on density and the viscous heating term does not. This causes the granular temperature to increase unphysically in regimes of very low density. The nature of this problem is not clear at all and it is not mentioned in the literature. The reason for this unphysical behavior might be the following. A realistic system of grains is not closed with respect to the energy of the grains, even if the material is confined within the domain. The collisions of grains with the wall are inelastic and hence kinetic energy of the molecules making up the grains is dissipated through the walls of the confining domain. This effect is not taken into account in our modeling of the granular temperature boundary conditions.

As far as we this case of very dilute flow is just not considered. We therefore leave the density out of the term as a regularization being aware how crude this approach is. In the regime where the kinetic model is validated, for example the shearflow we will introduce in Section 3.1, it does not make a big difference as 0.64 is close enough to 1 and the density more or less acts only as a factor in the dissipation term.

For the implementation all equations are scaled by the specific density of the grains $\hat{\rho}_g$ such that the grain density in all equations is replaced by a dimensionless volume fraction c. This has an effect on the coefficients and their units. In Equation (1.1a), the scaling only changes ρ to c. In Equation (1.1b), the scaling divides the dynamic viscosity η by $\hat{\rho}_g$. In the equation for granular temperature (1.7), the scaling affects η in the same way as in Equation (1.1b) as well as the heat conductivity λ. The constants λ_0 and η_0 given below are therefore divided by $\hat{\rho}_g$. For the ease of notation we continue to denote volume fraction by ρ in this section. We should always keep in mind that for the implementation and hence every numerical experiment the

1.3. MODELING THE DILUTE AND DENSE FLOW OF GRAINS

density is actually a volume fraction.

Let us introduce a few specific parameters of the model. First, in most cases of the application of the model the volume force f on the right hand side of equation (1.1b) is just the gravity which we will denote by g. This replaces in (1.1b) the term f by ρg.

Furthermore, the flow of granular media depends on the elasticity of the collisions of grains. We introduce the coefficient of restitution e which is between 1 and 0 for the elastic collisions with no dissipation of granular temperature and the fully inelastic collisions respectively. Clearly, the dissipation of granular temperature through grain collisions depends on e. We therefore introduce a dependency of the dissipation coefficient ε_0 on e based on [GD99]

$$\varepsilon_0 = \frac{8}{\sqrt{\pi} D_{\text{grain}}} \left(1 - e^2\right) \left(1 + \frac{3}{32} c^*\right), \quad c^* = \frac{32\left(1-e\right)\left(1-2e^2\right)}{81 - 17e + 30e^2\left(1-e\right)},$$

where D_{grain} is the grain diameter. This model assumes the particles to be hard spheres. The exact form of these very complicated relations is definitely a point of uncertainty and one has to make a compromise between the exactness of the relation and the number of parameters that have to be fitted for very complicated relations.

Let us finally provide the values of some of the constants involved. The values of λ_0, η_0, ρ_c, ρ_{co} and T_0 are

$$\lambda_0 = 0.00034 \frac{\text{m}^2}{\text{s}}, \quad \eta_0 = 0.00023 \frac{\text{m}^2}{\text{s}}, \quad \rho_c = 0.64, \quad \rho_{\text{co}} = 0.6, \quad T_0 = 0.5. \tag{1.16}$$

These values are again approximations, a vast amount of literature exists on formulas for them. But let us also plainly state that for the modeling of granular flow, the uncertainties start at much earlier points and it is questionable whether even more complicated relations for variables like ε_0 are reasonable. In our opinion the challenge lies in keeping the number of parameters that enter the model at a reasonable amount, see [Kol00]. As we mentioned earlier, the parameter ρ_{co} has been introduced in this work and its specific value is not clear.

Furthermore, for the implementation, the equation of state for the density as a function of pressure and temperature is needed

$$\rho(p,T) = \begin{cases} \rho_c \dfrac{p}{\rho_c T + p} &, p \le p_{co} \\ \rho_c \dfrac{p + \rho_{co} T_0}{\rho_c T + p + \rho_c T_0} &, p > p_{co} \end{cases}, \quad \text{with} \quad p_{co} = \rho_0 T g(\rho_{co}). \quad (1.17)$$

To appreciate the complexity of the model and to summarize all the relations previously derived we will provide the full system of equations used for the simulation of granular flow.

The full model of granular flow is

$$\partial_t(\rho) + \operatorname{div}(\rho \mathbf{u}) = 0, \tag{1.18a}$$

$$\partial_t(\rho \mathbf{u}) + \operatorname{div}(\rho \mathbf{u} \otimes \mathbf{u}) - \operatorname{div}(\tilde{\sigma}) = \rho \mathbf{g} - \operatorname{grad}(p), \tag{1.18b}$$

$$\partial_t(\rho T) + \operatorname{div}(\rho \mathbf{u} T) = \frac{2}{3}(\tilde{\sigma}:\tilde{\kappa} - \operatorname{div} \mathbf{q}) - \rho \varepsilon T, \tag{1.18c}$$

$$\tilde{\sigma} = \eta \tilde{\kappa}, \text{ with } \tilde{\kappa}_{ij} = \frac{\partial u_i}{\partial x_j}, \tag{1.18d}$$

with the relations

$$g(\rho) = \left(1 - \frac{\rho}{\rho_c}\right)^{-1}, \tag{1.18e}$$

$$\mathbf{q} = -(\lambda \operatorname{grad}(T)), \quad \text{with}$$

$$\lambda = \lambda_K \left(1 + \frac{p_Y}{p_K}\right), \quad \lambda_K = \lambda_0 \sqrt{T} g(\rho), \tag{1.18f}$$

$$p = p_K + p_Y, \quad \text{where}$$

$$p_K = T g(\rho) \rho, \quad p_Y = \Theta(\rho - \rho_{co}) \cdot T_0 \cdot (\rho - \rho_{co}) g(\rho), \tag{1.18g}$$

$$\eta = \eta_K \left(1 + \frac{p_Y}{p_K}\right), \quad \text{with } \eta_K = \eta_0 \sqrt{T} g(\rho), \tag{1.18h}$$

$$\varepsilon = \varepsilon_K \left(1 + \frac{p_Y}{p_K}\right), \quad \text{with } \varepsilon_K = \varepsilon_0 \rho^2 \sqrt{T} g(\rho). \tag{1.18i}$$

1.3. MODELING THE DILUTE AND DENSE FLOW OF GRAINS

Variable	Inflow $\partial\Omega_{in}$	Outflow $\partial\Omega_{out}$	Solid walls $\partial\Omega_{sw}$
Density ρ	$\rho = \rho_{in}$	-	$\frac{\partial \rho}{\partial \mathbf{n}} = 0$
Velocity $\mathbf{u}$	$\mathbf{u} = \mathbf{u}_{in}$	$\frac{\partial \mathbf{u}}{\partial \mathbf{n}} = 0$	$\mathbf{u} = 0$ (no-slip)
Temperature T	$\frac{\partial T}{\partial \mathbf{n}} = 0$	$\frac{\partial T}{\partial \mathbf{n}} = 0$	$\frac{\partial T}{\partial \mathbf{n}} = 0$

Table 1.1: Implemented boundary conditions for the granular flow model.

The units of the involved quantities after scaling with $\hat{\rho}_g$ are

$$[\rho] = 1, \quad [\mathbf{u}] = \frac{m}{s}, \quad [p] = \frac{m^2}{s^2}, \quad [\eta] = \frac{m^2}{s}, \quad [\varepsilon] = \frac{1}{s}, \quad [\lambda] = \frac{m^2}{s}.$$

Looking at System (1.18) in its full complexity, it should be clear that analytical methods of investigation are extremely challenging to say the least. We will see in Chapter 3 where we validate the model that we have to consider very simple cases, for example the flow in a periodic shearing cell to be able to learn at all about the analytical behavior of this system.

Boundary conditions

In Section 1.1.2 we have given general statements on boundary conditions for NSE-type systems. All the conditions derived there are given for characteristic variables. They are derived with well-posedness in mind. We must certainly follow the number of boundary conditions that should be provided for the 3-dimensional system. However, the exact form of boundary conditions is also governed by the physical modeling. We provide 5 boundary conditions on the inflow as well as on solid walls and 4 conditions on the outflow. Table 1.1 lists the specific conditions used in our implemented model. We can see in Table 1.1 that two very common types of boundary conditions are used. They are Dirichlet boundary conditions and homogeneous Neumann boundary conditions. For the former, the value of the unknown is given on the boundary and for the latter the derivative of the unknown in normal direction to the boundary is set to 0.

For granular flow, the issue of boundary conditions that mimic the physical behavior of grains on a boundary is complicated, see for example [MP00]. For each variable the boundary conditions given in Table 1.1 raise questions. Let us look at the boundary condition for velocity. Normal fluids remain attached to boundaries due to two effects, adhesion and internal pressure. This is different for granular material. Grains clearly do not just stick on the boundary. But they also do not just slip on the boundary without any friction. They can either stick, slide or roll on the boundary.

All cases attribute to different boundary conditions which have to be realized with a partial slip including friction. This however assumes that we can approximate the boundary better than just with cuboids. The scope of this is large and we do not treat the issue in this work.

Regarding the density, we will see in the shear flow experiment 3.2.1 that the it behaves differently near the boundary. So the assumption of the Neumann boundary condition is quite brave. Also for granular temperature we are probably getting away too easily by assuming a Neumann boundary condition. The temperature on walls is usually lower than in the interior of the domain for granular material.

Chapter 2

Algorithms

The aim of this chapter is to derive, present and investigate a method for computing approximate solutions of the granular flow model (1.18) from Section 1.3. It is a priory not clear how to approach the solution of such a system. It inhabits both compressible and nearly incompressible flow regimes. The constitutive relations are nonlinear and the density is a variable bounded by the maximum packing fraction of grains ρ_c. The flow is non-Newtonian meaning the viscosity depends on the pressure and the granular temperature.

We have to make a few general choices for the class of algorithms to consider. Our method will be a nonlinear fractional step method (NFSM) meaning that advancing the system in time is decoupled into substeps. We have already commented on the history of these decoupling methods in the introduction Chapter. In Section 2.1 we discuss these choices and argue why we think a nonlinear, implicit, pressure based fractional step method seems a good choice. A nonlinear method seems to be necessary because of the strongly nonlinear coupling of the constitutive relations and the differential equations in (1.18). The method should be implicit to escape strong time step restrictions posed by the high viscosity.

Before specific discussions on the algorithm we introduce a notation to describe the algorithms and specifically the discretizations in Section 2.2. We will introduce the notation for working on a collocated, cell-centered Finite Volume (FV) grid of cuboids. Also we will describe FV discretizations for the general terms that appear in any Navier-Stokes Equations (NSE)-type system without giving the full discretization of any system yet.

The development of our method will be preceded by Section 2.3 where we will give an introduction to linear fractional step methods (LFSMs) exemplified by a linear pressure correction algorithm (LPCA). Our nonlinear algorithm will then be derived in a similar way as the methods in this section.

To our knowledge, a pressure based fractional step method with a nonlinear pressure equation has not been developed so far. In Section 2.4 we will derive the nonlinear pressure algorithm (NPA) which takes into account the nonlinear nature of our granular flow model. The algorithm introduces a novel nonlinear pressure equation (NPE) which is the nonlinear equivalent of the linear pressure correction equation (LPCE) from Section 2.3.2.

Let us describe the steps towards a NPE in more detail. We first split up the continuous system into two steps which we call predictor and corrector. The predictor is an implicit, linearized solution of intermediate velocity for an old pressure value. The corrector is a coupled system of two equations. We want to derive the NPE from the fully discretized system. Therefore we discretize the predictor and the corrector in space. To derive the NPE from the discretized corrector, we make use of ideas from the derivation of LFSMs and [GBHL06, Section 4.1] but keep the full nonlinearities in the resulting system. Similar to the linear case, we use the mass conservation equation to arrive at an equation for the update of pressure, the NPE. It comes out as the nonlinear analog to the elliptic operator in the LPCE and other terms. Also like in the linear case, this is accompanied by an equation for the update of velocity. Unlike the linear case, the two equations remain coupled through a density upwind bias in the mass conservation. Hence, all novel aspects of our algorithm, a NPE and coupled pressure update and velocity update are both found in the corrector.

All the above discussions result in the presentation of the NPA for solving the complete time-dependent system (1.18) in Section 2.4.5. As the NPA involves the solution of the NPE, we need to solve a system of N nonlinear equations where N is the number of volumes. We use a variant of the Newton method for that purpose. Therefore we devote one section to the introduction of Newton methods and the specific method that we use. Following that, we write the NPE as a nonlinear equation $\mathcal{N}(p) = 0$ and provide its Jacobian $\mathcal{J}$. Also we give the detailed discretizations of both the nonlinear equation and the Jacobian.

Let us finally comment on Section 3.4 outside this chapter. For many aspects of the problem, the combination of a very complex system of equations (1.18) together with the partly nonlinear algorithm does not allow rigorous analysis. In the aforementioned Section 3.4 we provide a few numerical experiments which are aimed to support the derivation of the algorithm and investigate some of its properties in combination with the granular flow model.

2.1 Discussion on the type of numerical method

We are considering numerical solution methods or algorithms for the time-dependent NSE and ultimately our model for granular flow introduced in Section 1.3, System (1.18). Our aim is to solve that model for real processes of granular flow which in most cases means long intervals of time.

Different ways exist to categorize the existing algorithms for NSE-type systems. We choose to say that there are methods which treat the system in a coupled way and those that split up the process of obtaining a solution at a given time into substeps. The former usually discretize explicitly in time or treat at least most of the terms in an explicit fashion. This imposes problems as the viscosity (1.18h) in our granular flow model becomes very large for high volume fractions of grains.

In most cases the latter are implicit or semi-implicit methods that split one time step using multiple substeps. For the case of so-called projection methods one of the substeps is the solution of a second order partial differential equation (PDE) for the pressure or a correction of the pressure. We develop such a pressure based fractional step method with a nonlinear pressure equation. Let us explain why.

2.1.1 Arguments for an implicit, pressure based splitting

System (1.18) is of mixed type, as it is usual for NSE-type systems. This means it is neither always hyperbolic nor parabolic, rather it is of the class of incompletely parabolic PDE, compare for [Str76] and Section 1.1.2. Therefore it is a difficult task to solve the fully coupled system numerically where explicit schemes are usually preferred. For at least two reasons our System (1.18) is not suited for explicit solvers:

Diffusive time step restriction: Explicit schemes are bounded to a time step restriction of type $\tau < h^2/\eta$ by stability considerations, where h is the scale of the space discretization. This stability condition can be derived in different ways but is usually derived by von Neumann stability analysis. For further details see [Hir88, Section 8.3.1 Equation (8.3.18)] and for stability analysis using the matrix method see [Hir88, Section 10.3.1]. The condition is derived for the scalar heat diffusion equation discretized with explicit Euler time discretization and central finite difference space discretization.

For our much more complex system this consideration still suggests that explicit schemes can pose problems as the viscosity η in (1.18h) may become arbitrarily large for $\rho \to \rho_c$. Therefore we assume that in certain regimes we would be bound by this stability condition. Numerical experiments with explicit solvers show the relevance of this problem.

Algebraic restrictions in granular flow: With an explicit scheme, it is difficult to simultaneously fulfill the mass continuity (1.18a) and the constitutive relation (1.18g). Especially close to maximum packing of the granular media, the time step must become arbitrarily small to guarantee that one stays below the limit for ρ. One way to overcome this, would be to restrict the density. This however causes undesired loss of mass. Mass conservation however is a crucial property to our algorithm. Certainly the splitting approach we argue for also introduces an error but we consider the conservation of mass most critical.

Hence, we argue for an implicit scheme, where the common approach is to decouple the system. In the case of LPCAs or more generally LFSMs, the system is decoupled into the solution of the momentum conservation equation (1.1b) for velocity or momentum and an elliptic equation for the pressure, also sometimes called the pressure Poisson equation.

We follow this approach to derive an implicit pressure equation. Aside from the benefit of solving equations of which we know the type instead of solving a system of mixed type, our constitutive relations (1.18g) give another reason for making the pressure our unknown of choice.

Closely following the second argument above against an explicit scheme,

2.1. DISCUSSION ON THE TYPE OF NUMERICAL METHOD

let us look at the equation of state for the density as a function of pressure and granular temperature (1.17). We see that p may vary arbitrarily and nonetheless fulfills the constitutive relation (1.17). The density however is bound to finite limits between zero and the maximum packing fraction. This makes the pressure the logical choice as an unknown for numerical computations and rules out density based fractional step methods.

2.1.2 The need for a nonlinear method

Many attempts of finding a stable LFSM for granular flow have preceded this work. Our numerical experiments failed to successfully apply a LFSM to the granular flow model (1.18) and we are not aware of any work which has solved a similarely complex model using an implicit LFSM. LPCAs that work in both weakly compressible and incompressible regimes have been addressed extensively, see for example [Chu03] and [vVW03]. However, suggested by numerical experiments we found that the nonlinearity of our equations, especially the dependence of density on pressure does not allow the straightforward use of any of those schemes.

Let us explain this in more detail. Pressure based algorithms, especially the LPCAs for compressible flow where density depends on pressure, yield a term

$$\tau^{-1}(\rho(p^{n+1}) - \rho^n) \tag{2.1}$$

Usually an equation for a correction to the pressure $p' := p^{n+1} - p^n$ is desired so the term $\rho(p^{n+1})$ needs to be approximated in some way, for example by

$$\rho(p^{n+1}) \approx \rho^n + \frac{\partial \rho}{\partial p} p'. \tag{2.2}$$

In the case of a linear relation between density and pressure or simple nonlinear relations this may work fine. In our nonlinear case and especially for bounded density this approximation is wrong and causes problems.

First, it yields a pressure which corresponds to a linearly approximated density. This is a bad approximation but with a sufficient number of iterations might be cured. The more serious problem is that the density in (1.18) is bounded by the maximum packing volume fraction ρ_c by (1.12). We would have to make sure that any extrapolation of the density for the new pressure stays within the limiting density. We cure this problem by solving an equation for pressure, instead of pressure correction and taking into account the full nonlinearity of the problem. We will write the pressure equation with the full term (2.1) instead of approximating it. In this way the pressure equation inherits the density limit implicitly.

Furthermore we will introduce another type of nonlinearity in the NPE through the upwind discretization in the mass conservation equation. This is discussed shortly in [GBHL06]. We make use of some of the ideas therein, but modify them to the case of our NPE.

So we are using the full nonlinear dependence $\rho(p)$ from (1.17). The careful reader should recognize that by writing this, we have already simplified the relation and silently removed the dependence on the granular temperature. Let us discuss this in the next section.

2.1.3 The treatment of the temperature equation

In the first chapter and especially in Section 1.3 we have stated that the granular temperature is a key concept to a hydrodynamic model of granular flow. There is no such thing as isothermal granular flow. Hence in any model for granular flow, and specifically in System (1.18) we are dealing with a non-isothermal set of equations.

In the linear case, a temperature dependence would be incorporated into a pressure equation by modifying (2.2) to $\rho(p^{n+1}) \approx \rho^n + \frac{\partial \rho}{\partial p} p' + \frac{\partial \rho}{\partial E} E'$ where E is the energy of the system and E' is some approximate energy difference. In our case we have discussed in the previous section that we cannot approximate $\rho(p^{n+1})$ in a linear way. So for the case of a NPE this would add the complete equation for granular temperature to our last step in the algorithm, the corrector step in the derivation of the NPE in Section 2.4.3.

Another approach for the linear case that may be a hint also for granular temperature is that of [vVW03] of using the energy equation to construct a pressure correction equation. The energy equation for gases is a conservation equation. The equation for granular temperature however is a dissipative equation and it is not clear whether the concept of [vVW03] is even applicable in this case.

Finally we decide to derive our NFSM and the NPE in an isothermal fashion. This means that while solving for the pressure, we consider a constant granular temperature and only update it depending on density and pressure at the end of each time step. Certainly one may argue that the nonlinear dependence of pressure and density on temperature is not negligible. However, for the above reasons, the incorporation of the temperature equation into the pressure equation is out of reach.

2.2 General space discretization

Before continuing with the derivations of algorithms, we provide notations to describe the space discretizations used in this work. The notations are generic with respect to the dimension of the space. We first describe how the space is decomposed into a set of finite volumes, and then describe how continuous functions and operators are discretized on this grid. The grid is cell-centered which means that the discrete values of the unknowns are located in the center of the volume. The grid is collocated which means that the discrete values of all unknowns are are located at the same point for each volume. In our approach of using cuboids (for the 3D case) the boundary is approximated very roughly by the grid. We will not consider the issue of geometrically approximating boundaries for FV methods. This is treated in [Wes01, Chapter 11].

2.2.1 Notation

Let d be the space dimension. The computational domain Ω is decomposed into finite volumes of interval shape ($d = 1$), rectangular shape ($d = 2$), or cuboid shape ($d = 3$). These volumes are indexed in the canonical way by a

d-dimensional index

$$(i_1,\ldots,i_d) =: \mathbf{i} \in \mathbf{I}_d \subset \underbrace{\mathbb{N}_0 \times \cdots \times \mathbb{N}_0}_{d-\text{times}},$$

such that the index of a volume adjacent to the volume $\mathbf{i}$ in positive and negative Cartesian direction l is $(i_1,\ldots,i_l+1,\ldots,i_d)$ and $(i_1,\ldots,i_l-1,\ldots,i_d)$ respectively. For the purpose of a discrete formulation independent of the dimension and without restriction to cuboid domains we prefer the use of a one-dimensional lexicographical index. We introduce the bijective function

$$\pi_d : \mathbf{I}_d \mapsto J \subset \mathbb{N}_0 \qquad (2.3)$$

mapping every d-dimensional index in $\mathbf{I}_d$ to a one-dimensional index in J. Both ways of indexing will be used wherever they fit. Also, we omit the subscript d where it does not create confusion. The computational domain is discretized into $N := |J|$ finite volumes

$$\mathcal{V} := \{\mathbf{V}_{\pi(\mathbf{i})} | \mathbf{i} \in \mathbf{I}\}. \qquad (2.4)$$

We assume that the set of volumes $\{\mathbf{V}_j, j = 1,\ldots,N\} = \mathcal{V}$ is a nonoverlapping decomposition of Ω into subsets.

Let us now introduce the coordinates at the centers of the volumes, the centers of the faces and the centers of the neighbors of our volumes. The description of our discretization is local. Whenever we omit the lexicographical index as a subscript we mean the current volume denoted by j. Let

$$[x_1,\ldots,x_d] =: \mathbf{x} = \mathbf{x_i} = \mathbf{x}_{(i_1,\ldots,i_d)} = \mathbf{x}_{\pi^{-1}(j)} \in \mathbb{R}^d$$

be the coordinate of the node at the center of the volume $\mathbf{V}_j$ and in analog manner let

$$[h_1,\ldots,h_d] =: \mathbf{h} = \mathbf{h_i} = \mathbf{h}_{(i_1,\ldots,i_d)} = \mathbf{h}_{\pi^{-1}(j)} \in \mathbb{R}^d \qquad (2.5)$$

be the lengths of the volume as in Figure 2.1. We introduce for a lexicographical index j the shifted indexes

$$j \pm \mathbf{e}_l := \pi(\pi^{-1}(j) \pm \mathbf{e}_l) \quad \text{for} \quad \pi^{-1}(j) \pm \mathbf{e}_l \in \pi^{-1}(J), \qquad (2.6)$$

where e_l is the unit vector in the Cartesian direction l. Whenever it is convenient and does not create confusion we use a short form of the index $\pm e_l$. By writing $\pm$ we always mean both directions of one Cartesian direction l. This allows us to access neighbors in all Cartesian directions by a one-dimensional subscript. Each volume has $2d$ faces in the d Cartesian directions. Using the shifted indizes from (2.6) we denote a couple of faces in l-direction by $F_{\pm e_l}$ and the set of all faces of a volume by $\mathcal{F}_j := \{F_{\pm e_l} | l = 1, \ldots, d\}$ using the short form of the shifted index. The normals of the faces are denoted by

$$\mathbf{n}_{\pm \frac{1}{2} e_l}.$$

We define the set of all faces in Ω as a union

$$\mathcal{F} := \bigcup_{j \in J} \mathcal{F}_j. \tag{2.7}$$

To distinguish faces on the boundary and faces in the interior we introduce the sets

$$\bar{\mathcal{F}} := \{F \in \mathcal{F} | F \cap \partial \Omega \neq \emptyset\}, \tag{2.8a}$$
$$\mathring{\mathcal{F}} := \mathcal{F} \setminus \bar{\mathcal{F}}. \tag{2.8b}$$

Similar to (2.8) we introduce the following sets for volumes that have faces on the boundary and volumes in the interior.

$$\bar{\mathcal{V}} := \{\mathbf{V}_j \in \mathcal{V} | \exists F \in \mathcal{F}_j : F \in \bar{\mathcal{F}}\}, \tag{2.9a}$$
$$\mathring{\mathcal{V}} := \mathcal{V} \setminus \bar{\mathcal{V}}. \tag{2.9b}$$

using the one dimensional index j, the coordinates of the node centers and at the face centers are denoted by

$$\mathbf{x}_j \quad \text{and} \quad \mathbf{x}_{j \pm \frac{1}{2} e_l} := \mathbf{x}_j \pm \frac{h_l}{2} \mathbf{e}_l$$

respectively. We will however use the short form omitting j whenever this is reasonable and it is clear that we are speaking about an arbitrary but fixed volume. The notation and the decomposition of the domain Ω into the set of

CHAPTER 2. ALGORITHMS

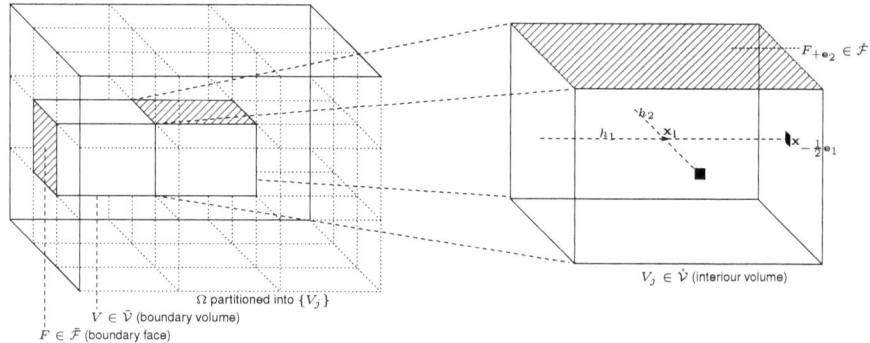

Figure 2.1: Visualization of a volume $V_j \in \overset{\circ}{\mathcal{V}}$, $j \in J$.

volumes $\mathcal{V}$ are visualized in Figure 2.1.

Note, at this point there is virtually no difference between a node at a boundary face and a node at an interior face, but we will refer to this in more detail later in this section and further in Section 2.5.1.

For an unknown q, being the pressure, a velocity component, etc. we introduce the following notation for values at center nodes, the face centers and the neighbor centers as

$$q_j := q(\mathbf{x}_j), \quad q_{j\pm\frac{1}{2}\mathbf{e}_l} := q(\mathbf{x}_{j\pm\frac{1}{2}\mathbf{e}_l}) \quad \text{and} \quad q_{j\pm\mathbf{e}_l} := q(\mathbf{x}_{j\pm\mathbf{e}_l})$$

respectively. For example, the vector value of velocity at the center of the east neighbor of the current volume j is denoted by $\mathbf{u}_{+\mathbf{e}_1}$.

This should give us enough notation to describe the discretization in an elegant way. We will describe the discretization in an arbitrary volume j.

2.2.2 Finite Volume space discretization

We use a cell-centered grid with collocated arrangement of the unknowns where all unknowns have values at the same node in the center of the volume. In this section we will not give the discretization of the complete equations (1.1) or (1.18). This will be done in later sections. The aim here is to provide the discretization in form of operators which are each responsible for the discretization of a certain type of PDE term.

These operators will then be used in later sections to write the discretiza-

2.2. GENERAL SPACE DISCRETIZATION

tion of various equations and systems in an elegant way. For the FV method all differential operators are integrated over volumes and most of them are transformed to surface integrals using the Gauss Theorem. For an introduction see [FP96, Section 4.1]. The surface of the volumes consists of the faces $\mathcal{F}$ from (2.8). So most of the differential operators in discrete form are written as discretization on faces. Since we store all unknowns at the volume centers, we introduce the interpolation between values at volume centers and values on faces for a face $F_{j \pm \mathbf{e}_l} \in \mathring{\mathcal{F}}$ of a volume with index j

$$q_{j \pm \frac{1}{2} \mathbf{e}_l} \approx \frac{1}{0.5(\mathbf{h}_j)_l + 0.5(\mathbf{h}_{j \pm \mathbf{e}_l})_l} \left((\mathbf{h}_{j \pm \mathbf{e}_l})_l q_j + (\mathbf{h}_j)_l q_{j \pm \mathbf{e}_l} \right). \tag{2.10}$$

This is certainly a very simple form of interpolation and many different and more sophisticated ways exist. For details on the approximation of face center values for FV methods see [FP96, Section 4.4]. The coefficients of System (1.18), for example the viscosity η, vary strongly between regimes of dilute and dense granular material but they are continuous. Certainly in the discrete case the jump in values between a volume with a high density of grains and one with a very low density may look like a discontinuity. But then Therefore we assume that this linear interpolation (2.10) should suffice. In

Also this interpolation will always be used when a scalar or vector valued solution of a certain FV discretized equation or system is needed on the face centers. Furthermore we will use another interpolation, the so-called upwinding introduced in [FP96, Section 4.4.1] for the discretization of certain operators.

Before we start with the space discretization let us make a remark on boundary conditions. We will postpone the implementation of boundary conditions to Section 2.5.1 because we want to give a general treatment for the linear discretization operators introduced in this section and the upcoming nonlinear equation discretizations. For now we just distinguish between faces that intersect with the boundary and faces on the interior of the domain. In the discretization the we will describe later we will use values on boundary faces without yet going into detail on how they are obtained.

Let us now consider the space discretization of PDEs. In the FV method we distinguish between two types of terms. These are surface integrals and volume integrals. The basics of FV discretization of these integrals are given in [FP96, Sections 4.2 and 4.3] respectively.

The models described in Chapter 1 contain terms like the divergence operator div, the gradient operator grad, concatenations of them as well as algebraic terms not involving differentials. Let q be a scalar field and let $\mathbf{v}$ be a vector field. In the following we call terms of the form $\mathrm{div}\,(\mathbf{v}q)$ *convective* terms. Those involving the concatenation of the div and grad operator in the form of $\mathrm{div}(\phi\,\mathrm{grad}(q))$ are called *diffusive* terms. To both we apply the Gauss Theorem when deriving a FV discretization. The resulting surface integrals are then approximated.

Convective terms

We start with the convective terms such as the second term in (1.1b) after applying the Gauss Theorem.

$$\int_{\partial \mathbf{V}} q\mathbf{v} \cdot \mathbf{n}\, dS. \tag{2.11}$$

Here q is a scalar field, for example density ρ and $\mathbf{v}$ is a vector field, for example the velocity $\mathbf{u}$. In all approaches to approximate the surface integral in (2.11) we start by summing the values on face centers over all faces of a volume. This is called the midpoint quadrature rule in the $3D$ form of

$$\int_{\partial \mathbf{V}} q\mathbf{v} \cdot \mathbf{n}\, dS \approx \sum_{F_{\pm \mathbf{e}_l} \in \mathcal{F}} \mathrm{area}(F_{\pm \mathbf{e}_l}) \left((\mathbf{v}q)_{\pm \frac{1}{2}\mathbf{e}_l}\right)_l \left(\mathbf{n}_{\pm \frac{1}{2}\mathbf{e}_l} \cdot \mathbf{e}_l\right). \tag{2.12}$$

The difference in the discretization results from the approximation of the midpoint or center face values. Let us introduce two different discretizations for the resulting terms. The first possibility is to approximate the face values by

2.2. GENERAL SPACE DISCRETIZATION

linear interpolation of the volume center values

$$\mathcal{DIV}(q,\mathbf{v}) := \sum_{F_{\pm e_l} \in \mathcal{F}} \text{area}(F_{\pm e_l}) \left((vq)_{\pm\frac{1}{2}e_l}\right)_l \left(\mathbf{n}_{\pm\frac{1}{2}e_l} \cdot \mathbf{e}_l\right), \text{ where} \quad (2.13a)$$

$$\left((vq)_{\pm\frac{1}{2}e_l}\right)_l = \begin{cases} (\mathbf{v}_{\pm\frac{1}{2}e_l})_l q_{\pm\frac{1}{2}e_l} & \text{using (2.10)}, F_{\pm e_l} \in \overset{\circ}{\mathcal{F}} \\ (\mathbf{v}_{\pm\frac{1}{2}e_l})_l q_{\pm\frac{1}{2}e_l} & \text{, otherwise} \end{cases}. \quad (2.13b)$$

As previously mentioned, the discussion on how to obtain the face values on the boundary is postponed to Section 2.5.1.

Let us additionally introduce a first order upwind discretization. The function q on the faces is interpolated in direction of the flow depending on some velocity. We will remark on the reasons for such a discretization in detail later in the derivation of the algorithm 2.2 in Section 2.4. So let us introduce a function which upwinds a quantity q depending on a velocity $\mathbf{u}$

$$UW(q, \mathbf{u}, F_{\pm e_l}) = \begin{cases} q & , \mathbf{u}_{\pm\frac{1}{2}e_l} \cdot \mathbf{n}_{\pm\frac{1}{2}e_l} \geq 0 \\ q_{uw} & , \mathbf{u}_{\pm\frac{1}{2}e_l} \cdot \mathbf{n}_{\pm\frac{1}{2}e_l} < 0 \end{cases}, \text{ with} \quad (2.14a)$$

$$q_{uw} := \begin{cases} q_{\pm e_l} & , F_{\pm e_l} \in \overset{\circ}{\mathcal{F}} \\ q_{\pm\frac{1}{2}e_l} & , \text{otherwise} \end{cases}, \quad (2.14b)$$

where $F_{\pm e_l}$ denotes any face of the couple of faces in l-direction on which the upwinded quantity should be determined. The function UW maps to an upwind approximation for face values with respect to the velocity $\mathbf{u}$. This means that the value interpolated onto the face center depends on the flow direction at the face center. If the flow is in direction of the face with respect to the volume center, then the value of the volume center is used. If the flow is in the opposite direction then the value of the neighboring volume center is used. The first order upwind discretization for terms of the form (2.11) is then given by

$$\mathcal{DIV}_{UW(\mathbf{u})}(q,\mathbf{v}) := \sum_{F_{\pm e_l} \in \mathcal{F}} \text{area}(F_{\pm e_l})(\mathbf{v}_{\pm \frac{1}{2}e_l})_l UW(q, \mathbf{u}, F_{\pm e_l}) \left(\mathbf{n}_{\pm \frac{1}{2}e_l} \cdot \mathbf{e}_l \right).$$
(2.15)

For the case of discretizing a convective term we will always use $\mathbf{u} \equiv \mathbf{v}$. The reason for a notation that allows the upwinding by a different vector field than the actual convecting velocity is the discretization of the System (1.18) in time which we will introduce in the later Section 2.4.1. There we will work with the velocity field at different values of time and will make full use of the notation introduced here.

Diffusive terms

The second major type of differential terms in any NSE-type system are diffusive terms such as the third term in (1.1b) which, compared to the transport properties of convective terms model the viscous aspects of flow. They include some diffusivity coefficient ϕ which may be the viscosity η from (1.18h). For such diffusive terms of the form

$$\int_{\partial \mathbf{V}} \phi \, \text{grad}\,(q) \cdot \mathbf{n} \, dS$$

we start with the same basic integration rule as in (2.12) except that our vector field is now a gradient of a scalar field. We introduce the discretization

$$\mathcal{D}(q, \phi) := \sum_{F_{\pm e_l} \in \mathcal{F}} \text{area}(F_{\pm e_l}) \phi(\mathbf{x}_{\pm \frac{1}{2}e_l}) \left((\text{grad}\, q)_{\pm \frac{1}{2}e_l} \right)_l \left(\mathbf{n}_{\pm \frac{1}{2}e_l} \cdot \mathbf{e}_l \right), \text{ with} \quad (2.16a)$$

$$(\text{grad}\, q_{\pm \frac{1}{2}e_l})_l \approx \begin{cases} \frac{1}{(\mathbf{x}_{\pm e_l})_l - x_l}(q_{\pm e_l} - q) & , F_{\pm e_l} \in \mathring{\mathcal{F}} \\ \frac{1}{\left(\mathbf{x}_{\pm \frac{1}{2} e_l}\right)_l - x_l}(q_{\pm \frac{1}{2}e_l} - q) & , \text{otherwise} \end{cases}, \quad (2.16b)$$

where ϕ will be the viscosity η as in (1.18h) or the thermal conductivity λ from (1.18f) which are both continuous functions.

In many applications of flow problems the diffusivity ϕ is not a function but a constant and the discretization may look different in many standard FV references. This is because the flow in these cases is assumed to be Newtonian which is true for many standard fluids. The granular flow model (1.18) is a non-Newtonian model and hence we need to be able to discretized diffusive terms with varying viscosity.

Other terms

Aside from those differential operators which can be transformed to surface integrals there are those which have to be treated as volume integrals. A common example is the force f in the momentum equation (1.1b).

Another is the pressure gradient found in all NSE-like systems. Here we need to approximate spatial gradients in a volume. We choose the approximation by differencing interpolated face center values across the volume and denote one component of the spatial gradient by

$$\frac{\partial q}{\partial x_l} \approx \mathcal{G}_\mathcal{F}(q)_l := \frac{1}{h_l}(q_{+\frac{1}{2}\mathbf{e}_l} - q_{-\frac{1}{2}\mathbf{e}_l}). \qquad (2.17)$$

The interpolation of the face center values is carried out as in (2.10). Then the gradient $\mathcal{G}_\mathcal{F}$ is the vector made up of the components $\mathcal{G}_{\mathcal{F}l}$ with $l = 1, \ldots, d$. Generally, volumetric integrals are approximated by multiplying the volume with the value at the volume center. With $V := \mathrm{volume}(\mathbf{V})$ we have

$$\int_\mathbf{V} q \, d\mathbf{V} \approx V \cdot q. \qquad (2.18)$$

The pressure gradient then will be approximated using (2.17) for q in (2.18).

Note that in the above discretizations we have omitted mixed derivatives as they appear in (1.1c). Their treatment in an implicit discretization is quite straightforward in the interior of the domain as in [Hir88, p191ff] but is challenging on the boundary.

In general our discretization operators are of low order. The upwind discretization (2.15) is crucial to stability, see [EGH00, Section 5.2.3], but restricts us in the given form to a first order space discretization. Certainly

FV discretizations of higher order exist [Fle91b, Section 9.3.2] but only make sense combined with a better approximation of the domain.

We have provided all the spatial discretization operators that will be needed for the discretization of the equations in the following sections. As previously mentioned, we will first split up the continuous system into substeps in time, discretize these subequations in space and then proceed with the derivation of a nonlinear pressure equation (NPE). This discretization in time will be a type of fractional step methods and hence we proceed with an introduction to these methods.

2.3 Introduction to fractional step methods

Although the focus of this work are NFSMs, the idea of LFSMs provides the basis for our development. We have mentioned before that we call all methods that split the advancing of the system in time into multiple steps LFSMs. In the third paragraph on page 8 we have provided references to origins of these methods. Fractional step methods for the instationary NSE have first been mentioned by Kim and Moin in [KM85]. Let us give a short introduction on the terminology and relationship between the different methods that split up the solution process. Then we will present a LPCA which falls into the category of LFSMs. This serves as an introduction to the development of our NPA.

2.3.1 The variants of fractional step methods

The pressure based NFSM NPA we will introduce is based on the very popular 'divide and conquer' approach initially introduced for solving the incompressible NSEs. The basic idea is to split the numerical treatment of the different operators and unknowns in the equations thus solving the initially difficult problem in relatively easier sub-steps. This approach has different names under different modifications: operator splitting, fractional step method or projection method. The terms 'splitting scheme' and 'projection scheme' partially overlap. It is generally accepted to use 'projection scheme' if the scheme is

2.3. INTRODUCTION TO FRACTIONAL STEP METHODS

based on a Helmholtz decomposition of the velocity meaning it can be generated by a scalar potential and a vector potential. These schemes usually require that the pressure equation is solved after the convection-diffusion step. Recently approaches are reported in the literature which bridge the different algorithm types and provide general formulations into which all the above methods can be categorized. We specifically mention the very recent publication [NA07]. They present a framework which theoretically allows the incorporation of nonlinear pressure equations at least for incompressible flow in [NA07, Equations (21)-(24)].

All the above mentioned schemes can also be considered 'splitting schemes' since the convection-diffusion part is split from the imposition of the incompressibility constraint. For a detailed overview of the state of the art of the variants of these schemes see [GS98, Min01, GMS06]. In the compressible case the idea remains but the continuity equation adds a compressibility term to the pressure equation.

Another general concept to describe these algorithms is to write them in the form of fractional step methods from [FP96, Section 7.4.1]. We show how this might be done by taking the simplest case, the explicit Euler time discretization and schematically split the System (1.1) into a fractional step method. We consider for the one dimensional case the following form.

$$\rho^{n+1} = \rho^n + \tau(C_{UW}(\rho\mathbf{u})) \quad \text{(2.19a)}$$
$$\mathbf{u}^{n+1} = \mathbf{u}^n + \tau(C_{UW}(\rho\mathbf{u}\mathbf{u}) + D(\mathbf{u}) + P) \quad \text{(2.19b)}$$

where C_{UW}, D, P represent the convective, diffusive and pressure terms respectively. We omit the time indizes on these terms on purpose as they depend on the specific method used. This can now be split in many ways. One approach is the splitting into

$$\mathbf{u}^{n+\frac{1}{3}} = \mathbf{u}^n + \tau C_{UW}(\rho^n \mathbf{u}^n \mathbf{u}^{n+\frac{1}{3}}) \quad \text{(2.20a)}$$
$$\mathbf{u}^{n+\frac{2}{3}} = \mathbf{u}^{n+\frac{1}{3}} + \tau D(\mathbf{u}^{n+\frac{2}{3}}) \quad \text{(2.20b)}$$
$$\mathbf{u}^{n+1} = \mathbf{u}^{n+\frac{2}{3}} + \tau P. \quad \text{(2.20c)}$$

In the third step (2.20c), P is the gradient of a quantity which obeys a Helmholtz type equation obtained by enforcing Equation (2.19a). Depending on the specifics of the splitting together with the way the split equations are enforced the method has different names, but the basic idea of LFSMs stays the same for all methods.

For the convergence analysis of splitting methods we refer to [GRS07, Theorem 8.21] and the preceding text in [GRS07, Section 8.3.4].

2.3.2 Linear pressure correction algorithm

We will exemplify the concept of LFSMs by this short section on a LPCA for the System (1.1) on our collocated, cell-centered grid. This concept will later be used for the derivation of the NPA.

As in the previous section we will split up the solution of a full time step into substeps. The final step will be formulated as an elliptic equation for an unknown correction to a pressure computed in the previous time step. Therefore this special LFSM is called pressure correction or Chorin projection method. The latter name originates from the incompressible case where the values of old pressure and velocity are projected onto the divergence free space to satisfy the incompressibility condition $\text{div}(\mathbf{u}) = 0$.

So let us assume that

$$\rho = \rho(p), \text{ with } \rho^{n+1} = \rho(p^n) + \frac{\partial \rho}{\partial p}(p^n) \cdot (p^{n+1} - p^n), \tag{2.21}$$

is a valid approximation of the compressibility condition and that the pressure p^n is known from the previous iteration, and η is constant. To obtain a first guess of the velocity in the current time step we solve the discretized integral form of the momentum equation (1.1b) in componentwise form

$$\frac{V}{\tau}\left[(\rho u)_l^{n+\frac{1}{2}} - (\rho u)_l^n\right] + \mathcal{DIV}_{UW(\mathbf{u}^n)}((\rho u)_l^{n+\frac{1}{2}}, \mathbf{u}^n) - \mathcal{D}((\rho^n)^{-1}(\rho u)_l^{n+\frac{1}{2}}, \eta)$$
$$+ V\,\mathcal{G}_{\mathcal{F}l}(p^n) = 0 \quad (2.22)$$

for $(\rho u)^{n+\frac{1}{2}}$ and $l = 1, \ldots, d$ where the convective term is linearized. Let us moreover assume that all other changes in velocity are induced by the gradient of the pressure correction, then we reach the following condition for

2.3. INTRODUCTION TO FRACTIONAL STEP METHODS

the corrected velocity and pressure

$$\frac{V}{\tau}\left[(\rho u)_l^{n+1} - (\rho u)_l^{n+\frac{1}{2}}\right] = -V\,\mathcal{G}_{\mathcal{F}l}(p^{n+1} - p^n), \quad l = 1,\ldots,d \Leftrightarrow$$
$$(\rho\mathbf{u})^{n+1} = (\rho\mathbf{u})^{n+\frac{1}{2}} - \tau\,\mathcal{G}_{\mathcal{F}}(p') \tag{2.23}$$

where $p' := p^{n+1} - p^n$. We apply the discrete divergence $\mathcal{DIV}(1,\cdot)$ to (2.23). Here $\mathbf{1}$ is a vector where each entry is 1. Using $\mathcal{DIV}(1,\rho\mathbf{u}) = \mathcal{DIV}(\rho,\mathbf{u})$, the discretized Equation (1.1a)

$$\frac{V}{\tau}\left[\rho^{n+1} - \rho^n\right] + \mathcal{DIV}(\rho^{n+1},\mathbf{u}^{n+1}) = 0$$

and (2.21), we transform (2.23) into the LPCE

$$-\frac{V}{\tau}\left[\frac{\partial\rho}{\partial p}(p^n)\cdot p'\right] = \mathcal{DIV}(\rho^{n+\frac{1}{2}},\mathbf{u}^{n+\frac{1}{2}}) - \tau\,\mathcal{DIV}(1,\mathcal{G}_{\mathcal{F}}(p')) \tag{2.24}$$

for the unknown p'. The relation $\rho(p)$ may be the ideal gas law (1.2).

In the incompressible case of $\frac{\partial\rho}{\partial p}(p^n) = 0$ Equation (2.24) is elliptic, the last term on the right hand side is a discretization of the Laplace operator. Otherwise it is a Helmholtz-type equation. The term on the left side is there because we solve a compressible or weakly compressible system. Hence we have transformed the solution of one time step for System (1.1) into two steps of solving the linearized velocity equation (2.22) and the solution of a second order PDE (2.24) with a Laplace operator. We summarize the steps into the LPCA given in a schematic way in Algorithm 2.1. For the case of

---
Algorithm 2.1: Linear pressure correction algorithm (one time step)

1. Solve linearized momentum prediction (2.22) for $(\rho\mathbf{u})^{n+\frac{1}{2}}$ with old pressure
2. Solve pressure correction equation (2.24) for p'
3. Correct momentum to $(\rho\mathbf{u})^{n+1}$ using (2.23)
4. $p^{n+1} = p^n + p'$
5. $\rho^{n+1} = \rho(p^{n+1})$

---

incompressible flow as modeled by (1.3) spurious oscillations in the pressure solution are known to occur for LPCAs on grids with collocated arrangement of the unknowns and our discretization of the pressure equation, see [Wes01, Chapter 6.3, p. 235ff.]. There it is shown that the discretization allows an oscillating pressure solution completely independent of the velocity. Often this problem is cured by the so called Rhie-Chow approach introduced in [RC83] which introduces a certain velocity interpolation preventing these oscillations.

We have found these sometimes called *checkerboard modes* when using Algorithm 2.1 for computing incompressible flow but we have chosen a different approach to cure the problem. Also for incompressible flow, Vabishchevich in [VPC96] and Chuiko and Lapanik in [CL05] apply a regularization operator to the right hand side of the pressure correction equation (2.24) to avoid pressure oscillations. This regularization operator is $(\Delta - \Delta_{\text{ext}})p^n$ where the subscript means an extended stencil. This, plainly speaking connects the discrete pressure and velocity solutions and hence avoids spurious nodes.

However, because LPCAs are not the focus or this work and our upcoming discretization of the NPE does not seem to allow these oscillations we will not go into further details on this issue.

2.4 A nonlinear pressure based algorithm

Based on the considerations in Section 2.1 we will use the space discretization operators from Section 2.2 and the concept of LFSMs introduced in Section 2.3 to develop a fractional step method with a NPE in this section. The method will be called NPA and its development is motivated by the need to solve System (1.18). However, the algorithm is developed in general for NSE-type systems with nonlinear coefficients or nonlinear constitutive relations.

We begin with splitting the continuous system in time into substeps which we will call predictor and corrector. The corrector will then be discretized in space and from that we derive a NPE coupled to an equation for a corrected velocity. As the NPE is nonlinear we will shortly mention the variant of the Newton method which we use to solve the NPE before giving its detailed discretization and that of the Jacobian.

2.4.1 Time discretization

Before we split up the time step into substeps, let us mention how partial time derivatives are treated. For discretization of such terms involving a partial time derivative, let q be any of the hydrodynamic quantities (density, one component of velocity or momentum or granular temperature) depending on time and space. For the partial time derivative of q, $\partial_t q$, we use implicit Euler discretization

$$\frac{1}{\tau_{n+1}}(q^{n+1} - q^n), \qquad (2.25)$$

where τ_{n+1} denotes the difference between the $(n+1)^{\text{st}}$ and the nth moments in time. We further split one time step of τ into different fractional steps. The fractional time steps will be denoted by superscripts of the form $n + \alpha$, $0 < \alpha < 1$ as in Section 2.3.

We have discussed in Section 2.1.3 that the temperature equation is treated decoupled from the system. We will derive the substeps by assuming a known granular temperature during one time step and solve the temperature equation (1.18c) at the end as a last substep.

The first fractional step will be the prediction of a velocity which we will denote by $\alpha = \frac{1}{2}$. This step is very similar to the linear case for System (1.1), compare with (2.22). There is a difference we would like to mention. We believe that instead of a prediction of momentum we should predict velocity because of the characteristics of System (1.18). We want to make the prediction step independent of strong and nonlinear dependence of density on pressure (1.17). This dependency has already been discussed in Section 2.1.1.

Prediction of an intermediate velocity

For the prediction of an intermediate velocity $\mathbf{u}^{n+\frac{1}{2}}$ we solve a linearized version of the velocity equation (1.18b)

$$\frac{\rho^n \mathbf{u}^{n+\frac{1}{2}} - \rho^n \mathbf{u}^n}{\tau} + \operatorname{div}\left(\rho^n \mathbf{u}^{n+\frac{1}{2}} \otimes \mathbf{u}^n\right) - \operatorname{div}\left(\eta(p^n)\kappa(\mathbf{u}^{n+\frac{1}{2}})\right) + \operatorname{grad} p^n - \rho^n \mathbf{g}^{n+1} = \mathbf{0}. \qquad (2.26)$$

Because the gravity is a constant vector, we always write $\mathbf{g}^{n+1}$ regardless of the fractional step in which it is used. It is important to note that we use the density from the old time step t^n in the prediction of velocity. The reason for this is that our final aim is the construction of a pressure equation. The dependence of $p(\rho)$ in (1.18g) is highly nonlinear, which means for a small error in density prediction a large error in the initial guess for the pressure is possible, especially in the dense regime.

Similar to LPCA from Section 2.3.2, the prediction of velocity is necessary because the convective and diffusive terms are not taken into account in the pressure equation. Only the change of velocity induced by the change in the pressure gradient is taken into account in the pressure equation.

There exist schemes that predict a density from the mass conservation equation to be used in the velocity prediction. Even though this is tempting we refrain from this practice again using the argument above that the dependence of density on pressure is very delicate and a small error in density corresponds to huge differences in the pressure. Furthermore, by the construction of the pressure equation we will see that the discrete version of (1.18a) will be satisfied exactly without a density prediction.

A system for an updated pressure

To construct the corrector, a nonlinear operator for the pressure together with an algebraic correction for velocity will be derived. Ideally, the corrected velocity $\mathbf{u}^{n+1}$ should satisfy, amongst mass continuity the fully nonlinear and implicitly time-discrete version of the velocity equation (1.18b)

$$\frac{\rho^{n+1}\mathbf{u}^{n+1} - \rho^n \mathbf{u}^n}{\tau} + \operatorname{div}\left(\rho^{n+1}\mathbf{u}^{n+1} \otimes \mathbf{u}^{n+1}\right) - \operatorname{div}\left(\eta(p^{n+1})\kappa(\mathbf{u}^{n+1})\right) \\ + \operatorname{grad} p^{n+1} - \rho^{n+1}\mathbf{g}^{n+1} = 0. \quad (2.27)$$

So far we have obtained velocity and pressure fields which satisfy the prediction equation (2.26). To obtain the operator for the difference between the two fields we subtract (2.26) from (2.27). The splitting happens when we drop the difference in velocity convection and diffusion. With this in mind, the subtraction gives equation (2.28a).

Speaking in the terms of Section 2.3.1 this effectively splits the solution method into a step for convection and diffusion using an old pressure field and, as we will see next, an equation for an update of pressure. This as-

2.4. A NONLINEAR PRESSURE BASED ALGORITHM

sumes that the major changes in the velocity are caused by the gradient of pressure.

Furthermore, both the corrected velocity and the density should satisfy the time-discrete version of the mass balance (1.18a), which gives equation (2.28b). These two requirements yield for the new pressure and velocity a coupled system of equations

$$\frac{\rho^{n+1}\mathbf{u}^{n+1} - \rho^n \mathbf{u}^{n+\frac{1}{2}}}{\tau} + \operatorname{grad}(p^{n+1} - p^n) - (\rho^{n+1} - \rho^n)\mathbf{g}^{n+1} = \mathbf{0}, \quad (2.28a)$$

$$\frac{\rho^{n+1} - \rho^n}{\tau} + \operatorname{div}(\rho^{n+1}\mathbf{u}^{n+1}) = 0, \quad (2.28b)$$

$$\rho^{n+1} = \rho(p^{n+1}). \quad (2.28c)$$

So we have split up the solution of System (1.18) into a prediction step (2.26) with the convective and diffusive terms and a step of enforcing the correct pressure and velocity fields (2.28). This is as far as we will go for the system which is still continuous in space. Hence the task preceding the derivation of a NPE will be the space discretization.

But before, let us look at how much explicitness we have introduced through our splitting. We do this by plugging our solution $\rho^n \mathbf{u}^{n+\frac{1}{2}}$ of Equation (2.26) into Equation (2.28). Hence we see that the combination of all fractional steps is effectively the solution of the system

$$\frac{\rho(p^{n+1})\mathbf{u}^{n+1} - \rho^n \mathbf{u}^n}{\tau} + \operatorname{div}(\rho^n \mathbf{u}^{n+\frac{1}{2}} \otimes \mathbf{u}^n) - \operatorname{div}(\eta(p^n)\kappa(\mathbf{u}^{n+\frac{1}{2}})) =$$
$$- \operatorname{grad} p^{n+1} + \rho(p^{n+1})\mathbf{g}^{n+1}, \quad (2.29a)$$

$$\frac{\rho(p^{n+1}) - \rho^n}{\tau} + \operatorname{div}(\rho(p^{n+1})\mathbf{u}^{n+1}) = 0 \quad (2.29b)$$

We see that the mass conservation is fulfilled for final fields with superscript $n+1$. However, the discretization of the velocity is only semi-implicit with respect to convection and diffusion.

Above we have introduced for the continuous case the basic approach for deriving a fractional step method for solving System (1.18). But we have not yet reached a satisfactory result. The System (2.28) is still difficult so solve in the given form and we have not yet derived a pressure equation. These steps towards the NPA will be done for the discretized equations only and our NPE will be solely a discrete concept. Therefore we will discretize the prediction step (2.26) and the coupled correction system (2.28) in space in the following section and then derive the NPE.

2.4.2 Spatial discretization of the split system

The equations involved in the split solution process are (2.26), (2.28) and an implicitly discretized, linearized version of (1.18c). We give a FV discretization of these equations using the space discretizations provided in Section 2.2. This will then be used to derive the NPE from the discrete equations.

So we first integrate the equations. For a control volume $\mathbf{V}$ fixed in time and space, the integral of (2.26) is

$$\int_{\mathbf{V}} \left[\frac{\rho^n u_l^{n+\frac{1}{2}} - \rho^n u_l^n}{\tau} + \mathrm{div}\,(\rho^n u_l^{n+\frac{1}{2}} \mathbf{u}^n) - \mathrm{div}\,(\eta(p^n)\,\mathrm{grad}(u_l^{n+\frac{1}{2}})) + \frac{\partial}{\partial x_l} p^n \right. $$
$$\left. - \rho^n g_l^n \right] d\mathbf{V} = 0, \quad (2.30)$$

for $l = 1, \ldots, d$ (d being the space dimension). Next we look at the integral form of the corrector (2.28). First we replace ρ^{n+1} by (2.28c) in (2.28a) and (2.28b). Then we integrate both equations over the volume V which yields.

$$\int_{\mathbf{V}} \left[\frac{\rho(p^{n+1}) \mathbf{u}^{n+1} - \rho^n \mathbf{u}^{n+\frac{1}{2}}}{\tau} + \mathrm{grad}\,(p^{n+1} - p^n) - (\rho(p^{n+1}) - \rho^n) \mathbf{g}^{n+1} \right] d\mathbf{V} = 0,$$
$$(2.31\mathrm{a})$$

$$\int_{\mathbf{V}} \left[\frac{\rho(p^{n+1}) - \rho^n}{\tau} + \mathrm{div}\,(\rho(p^{n+1}) \mathbf{u}^{n+1}) \right] d\mathbf{V} = 0.$$
$$(2.31\mathrm{b})$$

We apply the Gauss theorem to the terms involving a divergence operator.

2.4. A NONLINEAR PRESSURE BASED ALGORITHM

This results in

$$\int_V \left[\frac{\rho^n u_l^{n+\frac{1}{2}} - \rho^n u_l^n}{\tau} + \partial_l p^n - \rho^n g_l^{n+1} \right] d\mathbf{V}$$
$$+ \int_{\partial V} \left[(\rho^n u_l^{n+\frac{1}{2}} \mathbf{u}^n - \eta(p^n) \operatorname{grad}(u_l^{n+\frac{1}{2}})) \cdot \mathbf{n} \right] dS = 0, \quad (2.32)$$

$l = 1 \ldots, d$ for the velocity prediction (2.30) and

$$\int_V \frac{\rho(p^{n+1}) - \rho^n}{\tau} d\mathbf{V} + \int_{\partial V} (\rho(p^{n+1}) \mathbf{u}^{n+1}) \cdot \mathbf{n} \, dS = 0 \quad (2.33)$$

for (2.31b).

Let us write these integrated equations in FV discretization. We make use of the discretization operators introduced for the various PDE terms in Section 2.2. Let $\rho(p)$ be the pressure dependent density as in Equation (1.17). We describe the discretization locally for any volume $\mathbf{V}_j$ and drop the index j in the following. The discrete version of the predictor Equation (2.32) is

$$\frac{V}{\tau} \left[\rho^n u_l^{n+\frac{1}{2}} - \rho^n u_l^n \right] + \mathcal{DIV}_{UW(\mathbf{u}^n)}(\rho^n u_l^{n+\frac{1}{2}}, \mathbf{u}^n) - \mathcal{D}(u_l^{n+\frac{1}{2}}, \eta(p^n))$$
$$+ V \mathcal{G}_{\mathcal{F}l}(p^n) - V \rho^n g_l^n = 0 \quad (2.34)$$

for $l = 1, \ldots, d$. The discrete versions of the equations for the corrector (2.31a) and (2.33) are

$$\frac{V}{\tau} \left[\rho(p^{n+1}) \mathbf{u}^{n+1} - \rho^n \mathbf{u}^{n+\frac{1}{2}} \right] + V \mathcal{G}_{\mathcal{F}}(p^{n+1} - p^n) - V(\rho(p^{n+1}) - \rho^n) \mathbf{g}^{n+1} = 0$$
(2.35a)

$$V \frac{\rho(p^{n+1}) - \rho^n}{\tau} + \mathcal{DIV}_{UW(\mathbf{u}^{n+1})}(\rho(p^{n+1}), \mathbf{u}^{n+1}) = 0.$$
(2.35b)

Note that we keep the full nonlinearities in the corrector in order to derive a NPE. The linearizations in the predictor serve the purpose of a low cost prediction.

Let us remark on the upwind discretization in Equation (2.35b). There are similar ideas introduced by Bijl, Wesseling and others for the linear case. They do not leave the pressure equation and velocity correction equation coupled but they discretize the mass conservation equation using a first order upwind scheme similar to (2.35b).

There seem to be two good reasons for this. First, Wesseling suggests in [vVW01, Section 4] that the application of what he calls a first order density upwind bias in the mass conservation equation is necessary for stability in compressible flow. This approach is introduced in [BW98, Section 3.2 and Equation (27)]. The second reason is that pressure oscillations are avoided by this approach as explained in [GBHL06, Remark, pg. 11].

We will not give special attention to the discretization of the temperature equation (1.18c) at this point because it is not used in the derivation of the NPE and its discretization does not yield any further insights. Let us just remark that the convection and diffusion parts are discretized as the convection and diffusion in (2.34). The dissipation term T_{diss} is implicitly treated as a volume term and the derivatives in the viscous heating term are explicitly discretized using (2.17).

2.4.3 Derivation of the nonlinear pressure equation

Our goal is to obtain a pressure operator with properties close to the Helmholtz operator for the LPCA from Section 2.3.2. In the linear case, this means applying the divergence operator to a corrector equation and making use of the continuity equation. We will follow a similar path by multiplying (2.35a) with $\left[\rho(p^{n+1})V\right]^{-1}$. This is permissible because by Equation (1.17) we know that $\rho(p) > 0$ at all times. The goal of this division by density is to be able to use the continuity equation (2.35b). This yields

$$\frac{1}{\tau}\left[\mathbf{u}^{n+1} - \rho^n\rho(p^{n+1})^{-1}\mathbf{u}^{n+\frac{1}{2}}\right] + \rho(p^{n+1})^{-1}\mathcal{G}_{\mathcal{F}}\left(p^{n+1} - p^n\right)$$
$$- (1 - \rho(p^{n+1})^{-1}\rho^n)\mathbf{g}^{n+1} = 0. \quad (2.36)$$

Now we apply the upwind divergence operator $\mathcal{DIV}_{UW(\mathbf{u}^{n+1})}(\rho(p^{n+1}), \cdot)$ to equation (2.36). In particular this yields a term

$$\mathcal{DIV}_{UW(\mathbf{u}^{n+1})}(\rho(p^{n+1}), \mathbf{u}^{n+1})$$

2.4. A NONLINEAR PRESSURE BASED ALGORITHM

which is replaced with $V\frac{\rho(p^{n+1})-\rho^n}{\tau}$ using (2.35b). We then define the operator

$$\mathcal{L}p := \mathcal{DIV}_{UW(\mathbf{u}^{n+1})}\left(\rho(p), \rho(p)^{-1}\mathcal{G}_{\mathcal{F}}(\cdot)\right), \tag{2.37}$$

which should shorten the presentation significantly. We will apply this operator to both p^{n+1} and p^n. Also note that the unknown velocity $\mathbf{u}^{n+1}$ is part of the operator for the upwinding interpolation. We choose this form for now. When we later give the full algorithm we will introduce a form using an inner iteration between the unknown pressure p^{n+1} and the unknown velocity $\mathbf{u}^{n+1}$.

Using this notation, applying the upwind divergence operator to Equation (2.36) results in the pressure equation

$$\begin{aligned}\mathcal{L}p^{n+1} - \frac{V}{\tau^2}\rho(p^{n+1}) = &\mathcal{L}p^n - \frac{V}{\tau^2}\rho(p^n) \\ &+ \frac{1}{\tau}\mathcal{DIV}_{UW(\mathbf{u}^{n+1})}\left(\rho(p^{n+1}), \rho(p^{n+1})^{-1}\rho^n\mathbf{u}^{n+\frac{1}{2}}\right) \\ &+ \mathcal{DIV}_{UW(\mathbf{u}^{n+1})}\left(\rho(p^{n+1}), (1-\rho^n\rho(p^{n+1})^{-1})\mathbf{g}^{n+1}\right).\end{aligned} \tag{2.38}$$

It may seem strange to apply upwind divergence to a pressure gradient field or even to a gravity field, but looking closely at the definition of operator (2.15) we see that only the values of ρ are affected by this upwinding. It could be described by discretizing the divergence of the pressure gradients multiplied with the upwinded density. Furthermore, this derivation is purely discrete. One should not attempt to assign a meaning other than purely mathematical transformations to this pressure equation. Equation (2.38) together with

$$\mathbf{u}^{n+1} = \left[\rho(p^{n+1})\right]^{-1}\left(\rho^n\mathbf{u}^{n+\frac{1}{2}} - \tau\mathcal{G}_{\mathcal{F}}(p^{n+1}-p^n) + \tau\left(\rho(p^{n+1})-\rho^n\right)\mathbf{g}^{n+1}\right), \tag{2.39}$$

obtained from (2.36) solve the corrector exactly. We have gained two advantages that Equations (2.38), (2.39) have over the System (2.35). The first is that Equation (2.38) with the operator $\mathcal{L}$ should have nice properties because it is the divergence of a gradient, though nonlinear. And secondly, with Equation (2.39) we have the basis for an iterative procedure to compute p^{n+1} and $\mathbf{u}^{n+1}$ only using scalar equations.

Before we come to this iterative procedure, let us note that we have arrived at a nonlinear equation for pressure, the NPE (2.38). In fact, in this discrete form, Equation (2.38) is a system of N scalar nonlinear equations where N is the number of volumes into which the domain is decomposed as in (2.4) and as such may be quite large. A method is needed to solve this large system. We choose a variant of the Newton method for this purpose. Therefore, before we proceed with the complete NPA we will shortly remark on variants of the Newton method for large systems of nonlinear equations.

2.4.4 The Newton method and its variants

The Newton method is originally a method for finding the roots of a nonlinear scalar function $f : I \to I$ where I is an interval $I \subset \mathbb{R}$. Let f' be the derivative of f with respect to x. Then, assuming that $f'(x) \neq 0$ in the whole domain and with certain assumptions on the initial value x^0, the sequence

$$x^{\alpha+1} = x^\alpha - \frac{f(x^\alpha)}{f'(x^\alpha)} \qquad (2.40)$$

converges towards a root of f. This idea can be extended to systems of nonlinear equations. For a detailed introduction to the Newton method and its variants see [QSS06, Chapter 6 and Section 7.1]. We give the equivalent of (2.40) for systems. Let $\mathcal{N}(q)$ be a system of N nonlinear functions for the discrete unknown q. Let further $\mathcal{J}$ be the Jacobian of $\mathcal{N}$ which contains the derivatives of all components of $\mathcal{N}$ with respect to all components of q and is a $N \times N$ matrix valued function. Then for a given q^0 the Newton method can be written as

$$q^{\alpha+1} = q^\alpha + \delta q^\alpha \quad \text{with } \delta q^\alpha \text{ from} \quad \mathcal{J}(q^\alpha)\delta q^\alpha = -\mathcal{N}(q^\alpha). \qquad (2.41)$$

Hence we see that if written in this way, each iteration of the Newton method involves the solution of a linear system. This is the point where the modern variants of the Newton method differ. All methods where the linear system involved in every step is not exactly solved are called quasi-Newton methods. The first type are the methods where Jacobian or its factorization for the linear solution method are not updated in every step. The second are the so called inexact or truncated Newton methods which we will use for our problem.

Truncated Netwon Methods solve the linear system in every step by an iterative procedure but the maximum number of linear iterations is fixed. The idea is that far away from the nonlinear solution, there is no point in solving the linear system in every step very precisely. The methods are then called according the linear iterative method. In our case, because the linear system will be solved by Krylov-subspace methods we are using a Newton-Krylov method as described in [QSS06, Section 7.1.2.2 pg. 290].

A robust implementation framework for using this class of truncated Newton methods for large nonlinear systems is provided by the PETSc, SNES software package [BGMS97]. We make use of their solvers and methods for the solution of our NPE.

Furthermore, in modern Newton methods a step control is usually applied. By step we mean the δq^k. This is to ensure that the step does not leave the region of quadratic convergence around the solution which exists for all these methods. Two methods are the most popular, the so call *line search* and the *trust region* step controls. Both methods use the direction from the computed Newton step but limit the length with which to go in that direction. The topic is very advanced and is out of the scope of this work. The book [QSS06] provides the basics very well. For a complete coverage of all modern developments on Newton methods including the step control we cite [Deu04, Chapters 2 and 3].

2.4.5 The nonlinear pressure algorithm

In the previous sections we have transformed System (2.35) into a system of the Equations (2.38) and (2.39). The two equations are still coupled through the upwinding velocity u. Therefore we will give in this section an algorithm for solving the latter two equations in an iterative procedure. First we will iterate in a nonlinear Gauss-Seidel fashion through the two Equations (2.38) and (2.39). For this procedure we need to introduce an extra iteration index.

Let us define another version of the pressure operator (2.37)

$$\mathcal{L}^{n+1,k+1} := \mathcal{DIV}_{UW(\mathbf{u}^{n+1,k})}\left(\rho(p^{n+1,k}), [\rho(p^{n+1,k+1})]^{-1}\mathcal{G}_{\mathcal{F}}(\cdot)\right),$$

where the first superscript denotes the time step and the second denotes the iterations between pressure equation and velocity correction equation. With this definition we introduce Algorithm 2.2 which shows the details of the method. With Algorithm 2.2 we have arrived at the complete procedure

---

Algorithm 2.2: Nonlinear pressure algorithm (one time step)

1 Initialize: $\rho^{n+1,0} = \rho^n$, $p^{n+1,0} = p^n$
2 Solve momentum prediction (2.34) for $\mathbf{u}^{n+\frac{1}{2},0}$
3 Set $\mathbf{u}^{n+1,0} = \mathbf{u}^{n+\frac{1}{2},0}$
4 **repeat**
5   Solve pressure equation for $p^{n+1,k+1}$ using the truncated Newton method

$$\mathcal{L}^{n+1,k+1} p^{n+1,k+1} = \mathcal{L}^{n+1,k+1} p^n + \frac{V}{\tau^2}\rho(p^{n+1,k+1}) - \frac{V}{\tau^2}\rho(p^n)$$
$$+ \frac{1}{\tau}\mathcal{DIV}_{UW(\mathbf{u}^{n+1,k})}\left(\rho(p^{n+1,k}), [\rho(p^{n+1,k+1})]^{-1}\rho^n\mathbf{u}^n\right)$$
$$+ \mathcal{DIV}_{UW(\mathbf{u}^{n+1,k})}\left(\rho(p^{n+1,k}), (1-\rho^n[\rho(p^{n+1,k+1})]^{-1})\mathbf{g}^{n+1}\right)$$

6   Correct velocity for $\mathbf{u}^{n+1,k+1}$ (using (2.39))

$$\mathbf{u}^{n+1,k+1} = [\rho(p^{n+1,k+1})]^{-1}\left(\rho^n\mathbf{u}^n - \tau\mathcal{G}_{\mathcal{F}}(p^{n+1,k+1} - p^n) + \tau(\rho(p^{n+1,k+1}) - \rho^n)\right)\mathbf{g}^{n+1}$$

7   Update old iteration values $(p,\mathbf{u})^{n+1,k} = (p,\mathbf{u})^{n+1,k+1}$
8 **until** $\|p^{n+1,k+1} - p^{n+1,k}\| < \epsilon$
9 Solve temperature equation for $T^{n+1}(p^{n+1}, \mathbf{u}^{n+1})$

---

for solving System (1.18). The description of the time discretization is fully given. We have written the space discretization in a very compact form which hides the details for the sake of presenting the time discretization, the NFSM NPA in an elegant way. We will follow up on this and write down the space discretization explicitly in the next Section.

As we have previously mentioned, we solve the NPE (2.38) by a variant of the Newton method, see Section 2.4.4. This method requires the evaluation of the Jacobian of the NPE. Therefore, in the next Section we will not only provide the discretization of the NPE but also find its derivative, the Jacobian

2.4.6 Detailed discretization

We provide a description of the discretization of the full NPE (2.38) as a nonlinear function $\mathcal{N}(p^{n+1,k+1})$ of the unknown discrete pressure. As we have discussed in Section 2.4.4 we use a truncated Newton method with step control to solve the NPE. The iteration within the Newton method, which we have denoted by α in Section 2.4.4 is the third nested iteration for which, out of notational convenience we do not give an index within this Section. We keep in mind that in each Newton iteration for the NPE the unknown is actually $p^{n+1,k+1,\alpha+1}$ and that the function $\mathcal{N}$ as well as the Jacobian are evaluated at the given pressure field $p^{n+1,k+1,\alpha}$. Remember that the first iteration index $n+1$ denotes the time step and the second $k+1$ denotes the iteration between (2.38) and (2.39). Because we want to solve a system as in Equation (2.41) we need to provide also the detailed form of the Jacobian $\mathcal{J}$.

In fact, $\mathcal{N}$ is a system of N equations and $\mathcal{J}$ is a $N \times N$ matrix valued function where N is the number of finite volumes. For the ease of notation we replace the discrete pressure $p^{n+1,k+1,\alpha}$ by p for this section. The gravity term is included in the NPE and used in the simulations. But to further shorten the presentation we omit it in the description of the discretization it is just another volume term and does not yield any further insights.

The discretization will be provided for arbitrary domains independent of the dimension. Without giving the specific implementation of the boundary conditions from Section 1.3.3 we will distinguish between faces that intersect with the boundary and faces in the interior of the domain. This is possible because the discretization of the function $\mathcal{N}$ is actually just a rule for an explicit evaluation of $\mathcal{N}(p)$ in every Newton step, see (2.41). To apply the Newton method, we write the equation from step 5 of Algorithm 2.2 in the form

$$\mathcal{N}^{n+1,k+1}(p) = 0, \quad \mathcal{N} : \mathbb{R}^N \mapsto \mathbb{R}^N,$$

which again is a system of N nonlinear equations where each equation represents the discretization of one volume $\mathbf{V}_j \in \mathcal{V}$.

The nonlinear function

The expression of $\mathcal{N}^{n+1,k+1}$ is very lengthy when written out fully. We try to give a description which does not omit any detail but we refrain from displaying the expression in a single equation. We split up each equation of $\mathcal{N}_j^{n+1,k+1}$ into five parts, $\mathcal{N}1_j$ to $\mathcal{N}5_j$.

$$\mathcal{N}_j^{n+1,k+1}(p) := + \underbrace{\mathcal{DIV}_{UW(\mathbf{u}^{n+1,k})}\left(\rho(p^{n+1,k}), \frac{1}{\rho(p)}\mathcal{G}_{\mathcal{F}}(p)\right)}_{=:\mathcal{N}1_j}$$
$$- \underbrace{\mathcal{DIV}_{UW(\mathbf{u}^{n+1,k})}\left(\rho(p^{n+1,k}), \frac{1}{\rho(p)}\mathcal{G}_{\mathcal{F}}(p^n)\right)}_{=:\mathcal{N}2_j}$$
$$+ \underbrace{\frac{V}{\tau^2}\rho(p)}_{=:\mathcal{N}3_j} - \underbrace{\frac{V}{\tau^2}\rho(p^n)}_{=:\mathcal{N}4_j} + \underbrace{\frac{1}{\tau}\mathcal{DIV}_{UW(\mathbf{u}^{n+1,k})}\left(\rho(p^{n+1,k}), \frac{1}{\rho(p)}\rho^n\mathbf{u}^n\right)}_{=:\mathcal{N}5_j}$$

(2.42)

such that

$$\mathcal{N}_j^{n+1,k+1}(p) = \mathcal{N}1_j(p) - \mathcal{N}2_j(p) + \mathcal{N}3_j(p) - \mathcal{N}4_j(p) + \mathcal{N}5_j(p). \quad (2.43)$$

Schematically comparing the NPE to a LPCE $\Delta p' = \text{div}\,(\rho \mathbf{u})^n$ then the term $\mathcal{N}5_j$ corresponds to the divergence of the old momentum. The term $\mathcal{N}1_j - \mathcal{N}2_j$ corresponds to the Laplacian of p' and the terms $\mathcal{N}3_j$ and $\mathcal{N}4_j$ appear because of the compressibility and nonlinearity of the problem.

Before we continue with the spatial discretization let us refer to Section 2.2. We use the discretization operators from there. Actually, the NPE (2.38) is already given in discretized form using these operators but certainly the schematic form and the abbreviation through the pressure operator $\mathcal{L}$ from (2.37) hide any details. This was useful to not distract from the derivation of the NPE.

For a detailed presentation of the space discretization, let $\mathbf{V}_j$ be any fixed volume in $\mathcal{V}$, see (2.4). We first use the discretization operator (2.15). At the end of Section 2.4.2 we have already remarked on the upwind discretization and the reasons for it. Because of the application of the upwind divergence

2.4. A NONLINEAR PRESSURE BASED ALGORITHM

operator to the full Equation (2.36) this operator is contained in many terms and therefore we introduce a factor $D_{\pm e_l}$

$$D_{\pm e_l} := \text{area}(F_{\pm e_l}) UW(\rho(p^{n+1,k}), \mathbf{u}^{n+1,k}, F_{\pm e_l})(\mathbf{n}_{\pm \frac{1}{2}e_l} \cdot \mathbf{e}_l) \qquad (2.44)$$

for using it in the sums over faces as in (2.12). We then write

$$\mathcal{N}1_j(p) = \sum_{\substack{F_{j\pm e_l} \\ l=1,\ldots,d}} D_{j\pm e_l} \left(\frac{1}{\rho(p)}\right)_{j\pm\frac{1}{2}e_l} \left[\mathcal{G}_{\mathcal{F}}(p)_{j\pm\frac{1}{2}e_l}\right]_l, \qquad (2.45a)$$

$$\mathcal{N}2_j(p) = \sum_{\substack{F_{j\pm e_l} \\ l=1,\ldots,d}} D_{j\pm e_l} \left(\frac{1}{\rho(p)}\right)_{j\pm\frac{1}{2}e_l} \left[\mathcal{G}_{\mathcal{F}}(p^n)_{j\pm\frac{1}{2}e_l}\right]_l, \qquad (2.45b)$$

$$\mathcal{N}3_j(p) = \frac{V}{\tau^2}\rho(p_j), \qquad (2.45c)$$

$$\mathcal{N}4_j(p) = \frac{V}{\tau^2}\rho(p_j^n), \qquad (2.45d)$$

$$\mathcal{N}5_j(p) = \sum_{\substack{F_{j\pm e_l} \\ l=1,\ldots,d}} D_{j\pm e_l} \left(\frac{1}{\rho(p)}\right)_{j\pm\frac{1}{2}e_l} \left[(\rho^n \mathbf{u}^n)_{j\pm\frac{1}{2}e_l}\right]_l. \qquad (2.45e)$$

where $\rho(p) := \rho(p, T^n)$ from (1.17).

Still some details are left for clarification in the above equations. These are the inverse pressure dependent density and the gradient of the unknown pressure on faces in $\mathcal{N}1, \mathcal{N}2$ and $\mathcal{N}5$. Using the linear interpolation (2.10), the former is given by

$$\left(\frac{1}{\rho(p)}\right)_{j\pm\frac{1}{2}e_l} = \begin{cases} \left(\frac{1}{\frac{1}{2}\mathbf{h}_l+\frac{1}{2}(\mathbf{h}_{j\pm e_l})_l}\left(\frac{1}{2}(\mathbf{h}_{j\pm e_l})_l\rho(p_j)+\frac{1}{2}\mathbf{h}_l\rho(p_{j\pm e_l})\right)\right)^{-1}, & F_{j\pm e_l} \in \overset{\circ}{\mathcal{F}} \\ \frac{1}{\rho(p_{j\pm\frac{1}{2}e_l})} & , \text{otherwise} \end{cases}$$
$$(2.46)$$

where $\rho(p_{j\pm\frac{1}{2}e_l})$ is the boundary value at the face $F_{j\pm e_l}$. The latter, the gradient of the unknown pressure on faces is a more involved issue. We need to approximate the term

$$\left[\mathcal{G}_{\mathcal{F}}(p)_{j\pm\frac{1}{2}e_l}\right]_l. \qquad (2.47)$$

To achieve a discretization which resembles the continuous case of applying divergence to the gradient we would need to use the same approximation for the gradient in both the momentum equation (2.34) and the pressure equation (2.38). The correct expression for the gradient would then be the average of gradients of the volume centers because the gradient in the momentum equation is computed in centers. This would yield

$$\frac{\frac{1}{2}\mathbf{h}_l\left[\mathcal{G}_\mathcal{F}(p)_{j\pm e_l}\right]_l + \frac{1}{2}(\mathbf{h}_{j\pm e_l})_l\left[\mathcal{G}_\mathcal{F}(p)_j\right]_l}{\frac{1}{2}\mathbf{h}_l + \frac{1}{2}(\mathbf{h}_{j\pm e_l})_l}.$$

This expression involves values of the pressure not only at neighboring volumes, but also at neighbors of neighbors. For a less complex discretization, we choose to approximate the pressure gradient in the pressure equation as

$$\left[\mathcal{G}_\mathcal{F}(p)_{j\pm\frac{1}{2}e_l}\right]_l \approx \begin{cases} \frac{p_{j\pm e_l}-p_j}{(\mathbf{x}_{j\pm e_l})_l-(\mathbf{x}_j)_l} & ,F_{j\pm e_l} \in \mathring{\mathcal{F}} \\ \frac{p_{j\pm\frac{1}{2}e_l}-p_j}{(\mathbf{x}_{j\pm\frac{1}{2}e_l})_l-(\mathbf{x}_j)_l} & ,\text{otherwise} \end{cases}. \quad (2.48)$$

We conclude that the nonlinear function $\mathcal{N}$ is given by (2.42) with (2.45) using (2.46) and (2.48). We have reached a stage where the discretization is given in enough detail to actually be implemented.

For using the Newton method we also need to provide a detailed discretization of the Jacobian matrix valued function for $\mathcal{N}$. This is done in the next Section.

The Jacobian

We will derive a Jacobian for the discrete system of equations $\mathcal{N}$ given in the previous Section. We will take the continuous partial derivatives of $\mathcal{N}$ and evaluate the resulting matrix valued function at the discrete values of pressure. So we define the Jacobian $\mathcal{J}$ as

$$\mathcal{J}_{j\beta}(p) := \frac{\partial \mathcal{N}_j^{n+1,k+1}}{\partial p_\beta^{n+1,k+1}}(p), \quad (2.49)$$

Before we proceed with the details of (2.49), let us remark on the general topic of computing Jacobians for Newton methods for large systems. We choose to find the analytical derivatives of the discrete equations and compute the exact value of the Jacobian $\mathcal{J}(p)$. For systems where this can be done with reasonable effort and where analytical derivatives exist for the discrete equations it is by far the most efficient method with regard to computation time. However, in any other case, there exist other approaches to assemble a Jacobian for such a complicated discretization. We want to give two examples without going into too much detail.

Automatic differentiation: This is a very promising and quite recent approach. In this approach, the system of nonlinear functions $\mathcal{N}$ is provided in a form where the differentials can be recognized by a certain library. This library then automatically finds the derivative of the functions and provides the code to evaluate these derivatives. Once the derivatives are found, the approach is equivalent to our approach. The difference is that we have to program the derivatives after we have calculated them which is not necessary with automatic differentiation. Recent publications in this area using the same library that we use for the Newton method are [HNS05] and [BH97].

Approximation by finite differences: This is the most obvious, straightforward, easiest but also slowest approach. The idea is to form the Jacobian by

$$\frac{\mathcal{N}(p + \delta p) - \mathcal{N}p}{\delta p}$$

or any more sophisticated discretization approach.

If implemented successfully, the first approach would be very tempting. There would be no further need to look at Jacobians, only the nonlinear function needs to be provided. The second approach is slow but very helpful for testing whether the Newton method is able to solve the nonlinear equation at all before having to derive a Jacobian.

Again, we choose to analytically determine $\mathcal{J}$ by forming the derivatives of $\mathcal{N}$ with respect to the unknown pressure p. This results in a very fast computation of the Jacobian.

The Jacobian has a band shape where only a few entries per line are nonzero. The exact form of a line of $\mathcal{J}$ depends on whether the volume associated with that line is in the interior of the domain or has faces on the boundary. For ease of notation we omit the superscript of $\mathcal{N}$.

$$\begin{pmatrix} \frac{\partial \mathcal{N}_0(p)}{\partial p_0} & \cdots & & & & & & \\ & \ddots & & & & & & \\ \cdots & \frac{\partial \mathcal{N}_j(p)}{\partial p_{j-e_2}} & \cdots & \frac{\partial \mathcal{N}_j(p)}{\partial p_{j-e_1}} & \frac{\partial \mathcal{N}_j(p)}{\partial p_j} & \frac{\partial \mathcal{N}_j(p)}{\partial p_{j-e_1}} & \cdots & \frac{\partial \mathcal{N}_j(p)}{\partial p_{j-e_2}} & \cdots \\ & & & \cdots & \frac{\partial \mathcal{N}_{j+1}(p)}{\partial p_{j+1}} & \cdots & & \\ & & & & \ddots & & & \\ & & & & & \cdots & \frac{\partial \mathcal{N}_N(p)}{\partial p_N} \end{pmatrix} \begin{matrix} \} V_0 \in \bar{\mathcal{V}} \\ \vdots \\ \} \mathbf{V}_j \in \overset{\circ}{\mathcal{V}} \\ \} V_{j+1} \in \bar{\mathcal{V}} \\ \vdots \\ \} V_N \in \bar{\mathcal{V}} \end{matrix}$$
(2.50)

In (2.50) we see a possible configuration of the Jacobian. Because we are considering arbitrary domains, it may happen that a line for a volume with faces on the boundary is inbetween lines representing volumes in the interior. Each line contains an entry $\frac{\partial \mathcal{N}_j}{\partial p_j}$ for the center node of the respective volume and a certain number of entries for neighboring nodes, depending on how many of the neighbors lie in the interior of the domain.

Let us derive $\mathcal{N}$ by the unknown pressure to find the specific form of $\mathcal{J}$. Again we split up $\mathcal{N}$ into $\mathcal{N}1$ to $\mathcal{N}5$ and take their derivatives separately. The part $\mathcal{N}4$ does not depend on the unknown pressure and hence does not contribute to the Jacobian. The simple term $\mathcal{N}3$ contributes only to the diagonal with the term

$$\frac{\partial \mathcal{N}3_j}{\partial p_j}(p) = \frac{V}{\tau^2} \frac{\partial \rho}{\partial p}(p_j). \qquad (2.51)$$

We have given $\rho(p)$ for the granular flow model in (1.17). The derivative of this relation is

$$\frac{\partial \rho}{\partial p}(p)\Big|_{T=T^n} = \begin{cases} \dfrac{\rho_c^2 T^n}{(p + \rho_c T^n)^2} & , p \leq p_{\text{co}} \\ \dfrac{\rho_c(\rho_c T^n + \rho_c T_0 - \rho_{\text{co}} T_0)}{(p + \rho_c T^n + \rho_c T_0)^2} & , p > p_{\text{co}} \end{cases}. \qquad (2.52)$$

Now let us look at the more complicated terms $\mathcal{N}1, \mathcal{N}2$ and $\mathcal{N}5$. Taking the

2.4. A NONLINEAR PRESSURE BASED ALGORITHM

derivative of these three terms is very similar between them so we give the derivative of all three together.

Let $\hat{p} \in \{p_j\} \cup \{p_{j \pm e_l}, l = 1, \ldots, d\}$ be either the unknown pressure in the volume center or in the center of any of the neighbors. Using the prefactor $D_{\pm e_l}$ introduced in (2.44) the derivatives of (2.45a), (2.45b) and (2.45e) can be written as

$$\frac{\partial \mathcal{N}1_j}{\partial \hat{p}}(p) = \sum_{\substack{F_{j \pm e_l} \\ l=1,\ldots,d}} D_{j \pm e_l} \left(\frac{\partial}{\partial \hat{p}} \left(\frac{1}{\rho(p)} \right)_{j \pm \frac{1}{2} e_l} \left[\mathcal{G}_\mathcal{F}(p)_{j \pm \frac{1}{2} e_l} \right]_l + \right.$$
$$\left. \left(\frac{1}{\rho(p)} \right)_{j \pm \frac{1}{2} e_l} \frac{\partial}{\partial \hat{p}} \left[\mathcal{G}_\mathcal{F}(p)_{j \pm \frac{1}{2} e_l} \right]_l \right), \quad (2.53)$$

$$\frac{\partial \mathcal{N}2_j}{\partial \hat{p}}(p) = \sum_{\substack{F_{j \pm e_l} \\ l=1,\ldots,d}} D_{j \pm e_l} \left[\mathcal{G}_\mathcal{F}(p^n)_{j \pm \frac{1}{2} e_l} \right]_l \frac{\partial}{\partial \hat{p}} \left(\frac{1}{\rho(p)} \right)_{j \pm \frac{1}{2} e_l}, \quad (2.54)$$

$$\frac{\partial \mathcal{N}5_j}{\partial \hat{p}}(p) = \sum_{\substack{F_{j \pm e_l} \\ l=1,\ldots,d}} D_{j \pm e_l} \left[(\rho^n \mathbf{u}^n)_{j \pm \frac{1}{2} e_l} \right]_l \frac{\partial}{\partial \hat{p}} \left(\frac{1}{\rho(p)} \right)_{j \pm \frac{1}{2} e_l}. \quad (2.55)$$

The above is still a general form of the derivative. Some of the parts need special attention if a face intersects with the boundary. Let us start with the derivative of the inverse pressure dependent wall density (2.46) which is essential to the derivatives (2.53)-(2.55). For the derivative with respect to the volume center and the neighbors we have

$$\frac{\partial}{\partial p_j} \left(\frac{1}{\rho(p)} \right)_{j \pm \frac{1}{2} e_l} = \frac{1}{\frac{1}{2} \mathbf{h}_l + \frac{1}{2} (\mathbf{h}_{j \pm e_l})_l} (\mathbf{h}_{j \pm e_l})_l \frac{\partial \rho}{\partial p}(p_j) \left(\rho(p)_{j \pm \frac{1}{2} e_l} \right)^2 \quad (2.56a)$$

$$\frac{\partial}{\partial p_{j \pm e_l}} \left(\frac{1}{\rho(p)} \right)_{j \pm \frac{1}{2} e_l} = \frac{1}{\frac{1}{2} \mathbf{h}_l + \frac{1}{2} (\mathbf{h}_{j \pm e_l})_l} (\mathbf{h}_l \frac{\partial \rho}{\partial p}(p_{j \pm e_l}) \left(\rho(p)_{j \pm \frac{1}{2} e_l} \right)^2 \quad (2.56b)$$

respectively for $F_{j \pm e_l} \in \mathring{\mathcal{F}}$. Otherwise the derivatives are zero by (2.46). Using (2.10) and (2.52) we then have fully derived the inverse pressure dependent wall density.

The last terms left for clarification are the derivatives of the pressure gradient faces that appear in (2.53). The derivative with respect to the volume center pressure p_j is given by

$$\frac{\partial}{\partial p_j}\left[\mathcal{G}_{\mathcal{F}}(p)_{j\pm\frac{1}{2}\mathbf{e}_l}\right]_l = \frac{-1}{(\mathbf{x}_{j\pm\mathbf{e}_l})_l - (\mathbf{x}_j)_l}.$$

The derivatives with respect to the neighboring volume centers are given by

$$\frac{\partial}{\partial p_{j\pm\mathbf{e}_l}}\left[\mathcal{G}_{\mathcal{F}}(p)_{j\pm\frac{1}{2}\mathbf{e}_l}\right]_l = \begin{cases} \frac{-1}{(\mathbf{x}_{j\pm\mathbf{e}_l})_l - (\mathbf{x}_j)_l} & , F_{j\pm\mathbf{e}_l} \in \mathring{\mathcal{F}} \\ 0, \text{otherwise} \end{cases}. \quad (2.57)$$

Let us finally conclude how the Jacobian looks. The general shape of the matrix valued function is given in (2.50). There we see that it has a band shape very similar to the discretization of a PDE where each line represents the discretization of one finite volume. The difference for the Jacobian is that the columns contain derivatives with respect to neighbor volumes instead of the coefficients with respect to neighbor volumes as in PDE discretization. The number of entries in each line depends on how many faces of the volume for this line intersect with the boundary.

The major terms on the diagonal of the Jacobian are (2.51) and (2.53). They are the derivatives of the compressibility term and the pressure operator (2.37) respectively.

Certainly we have presented here only one possible way to discretize $\mathcal{N}$ and hence to compute the Jacobian $\mathcal{J}$. But as we are using a truncated Netwon Method where the linear system involving the Jacobian as in (2.41) is only solved approximately it is not critical to have an exact Jacobian. The reason why we find the derivative analytically is computation efficiency.

Furthermore we have not given a detailed treatment of the discretization of boundary conditions. Even though in Equations (2.46),(2.48),(2.56) and (2.57) we have distinguished between faces that intersect with the boundary and those in the interior we have only made use of values on the faces. We postpone the question of how these values are obtained to Section 2.5.1.

2.5 Initial and boundary conditions

Up to this point we have given an overview of the theory of boundary conditions for NSE-type systems in Section 1.1.2 and we have mentioned the number and type of continuous boundary conditions and the problems that exist for the granular flow model in Section 1.3.3 and especially in Table 1.1.

In the discretization of the NPE and its Jacobian in Section 2.4.6 we have distinguished whether a discretization point is on a face that intersects with the boundary or on an interior face in Equations (2.46),(2.48),(2.56) and (2.57). However, we have not talked about how the values at boundary faces are approximated. The issue of initial conditions has been completely left out so far.

In this Section we will focus on how the boundary conditions enter the discretization and how initial conditions for the granular flow model (1.18) can be chosen. We are aware that we treat the issue of boundary and initial conditions in a very narrow scope as this is not the focus of this work. Especially for granular flow, the topic is vast and complicated as we have hinted in Section 1.3.3 and well deserves treatment in a work of its own.

2.5.1 Approximation of boundary conditions

We have already mentioned in Section 1.3.3 and particularly in Table 1.1 that we consider two types of boundary conditions. Let us repeat them. The first are Dirichlet conditions of the form

$$q = \bar{q} \quad \text{on} \quad \partial\Omega \qquad (2.58)$$

where q is any unknown density, pressure, velocity component etc. and $\bar{q}$ is a given function on the boundary. The second are homogeneous Neumann boundary conditions where the derivative of an unknown on the boundary and in normal direction of the boundary is set to 0

$$\frac{\partial q}{\partial \mathbf{n}} = 0 \quad \text{on} \quad \partial\Omega. \qquad (2.59)$$

Here $\mathbf{n}$ is the normal on the boundary.

We distinguish between two cases of how these boundary conditions are implemented. The first case is the discretization of the NPE in Section 2.4.6. This is effectively an explicit discretization because as we have mentioned in the first paragraph of Section 2.4.6 the function $\mathcal{N}$ is just evaluated at the given pressure field $p^{n+1,k+1,\alpha}$ in every Newton step.

The second case are the linear implicit discretizations as for example the operators $\mathcal{DIV}$ and $\mathcal{D}$ from (2.13) and (2.16) respectively. They are used for example in the discretization (2.34) of the velocity prediction (2.26) but also in the discretization of the temperature equation.

Let $\bar{j}$ be the index of a volume in $\bar{V}$ as in (2.9) and assume without loss of generality and for ease of explanation within this Section that the face $F_{+e_1} \in \mathcal{F}_{\bar{j}}$ is in $\bar{\mathcal{F}}$ from (2.7). Let us first consider the case of Dirichlet boundary conditions.

Dirichlet boundary conditions: There is no difference in their implementation between the NPE and the operators $\mathcal{DIV}$ and $\mathcal{G}_\mathcal{F}$. So for any unknown q, let $\bar{q}$ be the Dirichlet condition on the boundary. Then for the linear implicit discretization operators in (2.13b), (2.14b), (2.16b), (2.17) and for the discretization of the NPE in (2.44), (2.46), (2.48) we set

$$q_{\bar{j}+\frac{1}{2}e_1} = \bar{q}(\mathbf{x}_{\bar{j}+\frac{1}{2}e_1}). \tag{2.60}$$

Note that for the NPE the unknown is always called p instead of the general q because p is in that case the only unknown used in the discretization.

Homogeneous Neumann boundary conditions: They are treated differently for the implicit linear operators and the NPE. In the discretization operator $\mathcal{D}$ for diffusive terms in Equation (2.16) we set

$$(\operatorname{grad} q_{\bar{j}+\frac{1}{2}e_1})_1 = 0$$

in (2.16b). In similar fashion for the pressure operator (2.37) of the NPE (2.38) we set

$$\left[\mathcal{G}_\mathcal{F}(p)_{\bar{j}+\frac{1}{2}e_1}\right]_1 = 0$$

2.5. INITIAL AND BOUNDARY CONDITIONS

in Equation (2.48).

For the discrete divergence operator $\mathcal{DIV}$ in (2.13b), (2.14b), the operator $\mathcal{G}_\mathcal{F}$ in (2.17) and the respective parts of the NPE discretization in (2.44), (2.46) we set

$$q_{\bar{j}+\frac{1}{2}e_1} = q_{\bar{j}}$$

to approximate the condition (2.59).

A special case are periodic boundaries. We will show how the periodicity is approximated in the discretization.

Periodic boundary conditions: Plainly speaking, the periodicity of a domain is realized by discretizing across the periodic boundary assuming the volume in the opposite direction of the periodic boundary to be the direct neighbor. This means that with respect to the discretization, a volume on the periodic boundary is discretized exactly like one on the interior with a different neighbor.

Let us exemplify this for the 1D case on the domain $[0, 1]$ which we decompose into 5 volumes $\mathbf{V}_1, \ldots, \mathbf{V}_5$ of length $h_1 = \frac{1}{5}$ each. If for example we look at the incompressible case of Equation (2.24) we use the discretization operator $\mathcal{D}$ from Definition (2.16) as $\mathcal{D}(q, 1)$ on an unknown q on this domain and assume the domain to be periodic, then the discretization in the volume $\mathbf{V}_1$ on the left periodic boundary looks like

$$-\frac{1}{(\mathbf{x}_5 - \mathbf{x}_{5+\frac{1}{2}e_1})_1 + (\mathbf{x}_{1-\frac{1}{2}e_1} - \mathbf{x}_1)_1}(q_5 - q_1) + \frac{1}{(\mathbf{x}_2)_1 - (\mathbf{x}_1)_1}(q_2 - q_1). \quad (2.61)$$

The notation seems a bit overly complicated for such a simple discretization but we want to remain as close as possible to our general notation used for the definition of $\mathcal{D}$ in (2.16). The difference between Equation (2.61) and (2.16) is in the denominator of the first term in (2.61) where we determine the length between the periodically connected volumes $\mathbf{V}_1$ and $\mathbf{V}_5$.

This discretization inhabits a problem. The resulting linear system to solve

$\mathcal{D}(q,1) = 0$ with the above periodic discretization is

$$\frac{1}{5}\begin{pmatrix} -2 & 1 & & & 1 \\ 1 & -2 & 1 & & \\ & 1 & -2 & 1 & \\ & & 1 & -2 & 1 \\ 1 & & & 1 & -2 \end{pmatrix} \cdot \begin{pmatrix} q_1 \\ q_2 \\ q_3 \\ q_4 \\ q_5 \end{pmatrix} = 0,$$

which has only rank 4 and is hence not invertible. This can be cured by fixing the value of the unknown at exactly one point of the periodic boundary. For example for the left boundary point, this removes the entry 1 at the upper right corner of the matrix and hence it has rank 5.

Here the 1D case looses its practicability because by setting one of the two boundary points we loose periodicity. But in the 2D or 3D case this a successful approach which we use.

2.5.2 Initial conditions for the granular flow model

We have found during the development of the NPA 2.2 that the choice of initial conditions is crucial for the convergence and has a strong impact on the maximum possible time step. As long as we are only dealing with an inflow into an empty domain the issue is quite clear. We set the density variable to a very low value of volume fraction, say 1×10^{-3}, we provide a low granular temperature of the same order which corresponds to very little movement of the few grains that are in the domain and use the relation (1.18g) to compute the corresponding pressure field. The initial state is then fully determined.

The topic becomes much more complicated if we start with an initial bulk of granular material which fills the domain. We have such a case for example in the shearflow experiment in Section 3.1 and in the falling block of grains in Section 3.4.1.

Generally speaking, initial conditions should be chosen with a given pressure profile and a given volume fraction as this determines the initial grain configuration. From these two fields a granular temperature should be computed again using (1.18g). In the case of simulations under the force of gravity, the best initial guess we have for the pressure without solving the system is a hydrostatic pressure profile where the pressure balances out the gravity in direction of the filling height H as $p = \rho g_H H$. By filling height we mean the distance the grains have from the bottom in a pile.

But even with this condition remains the problem that we introduce a discrete discontinuity in the density and hence in the pressure and the temperature between the region which is filled with grains and the region without grains. The value of ρ as a volume fraction then changes from below 0.64 to $1e-3$ from one volume to the next.

We have found that the algorithm NPA 2.2 together with our FV discretization is able to handle at least the hydrostatic initial conditions or the initially empty domain. It is however not clear how close to reality the time-dependent process is when we start, say the emptying of a Silo, with hydrostatic initial conditions. We observe that the granular material first compacts to a certain resting state before actual flow occurs. These problems are outside the scope of this thesis and should be treated in a work of its own.

Chapter 3

Validation and numerical simulations

In this chapter we present numerical experiments to validate, investigate and apply both the granular flow model introduced in Section 1.3 and the nonlinear pressure algorithm (NPA) introduced in Section 2.4. We start by presenting our main validation experiment, the granular shear flow. This is the best studied granular flow experiment and is approached analytically in [BLS$^+$01]. It serves as a basis for validating the model as well as the algorithm. Therefore we introduce the experiment in Section 3.1 and discuss its numerical setup in general. The results of the experiment with respect to various validation aspects will be given in the subsequent sections.

In Section 3.2 we will show that the presented hybrid model for granular flow can be solved with our numerical algorithm and is able to reproduce typical granular flow patterns. We will concentrate on the dense regime, since our model is equivalent to the often tested kinetic theory in dilute and intermediately dense flow. In Section 3.3 we will investigate the NPA for both Newtonian and non-Newtonian flow.

Section 3.4 will show investigations of properties of the NPA in combination with the granular flow model. This aims to support statements during the derivation of the algorithm and the model and should give some insights into the challenges posed by this combination and the properties of the NPA. Finally, in Section 3.5 we show the applicability of our approach to two selected industrial problems.

3.1 The granular shear flow experiment

We simulate the shearing experiment from the experimental and analytical work of Bocquet et al. in [BLS$^+$01]. In the experimental setup displayed in [BLS$^+$01, Figure 1, pg. 4] the granular media is sheared in a Couette geometry. The grains are agitated by the flow of air from the bottom. An inner cylinder rotates inside a resting outer cylinder. Inbetween the two cylinders the grains are constrained to a 12mm gap. In [BLS$^+$01, Section 4] it is stated that the behavior of the inner cylinder is very similar to the dynamics of a rough plate sliding across a granular layer. Therefore we carry out the simulation with exactly this configuration using periodic boundaries in the shearing direction.

Our domain is a plate as in Figure 3.1 which is 1.8mm thick, 12mm wide and 36mm long. The boundary conditions in x-direction are periodic to simulate the cylindrical layout in the experiment. The main difference between the experimental setup in [BLS$^+$01, Figure 1, pg. 4] and our setup is the flow of air. As the flow of air in [BLS$^+$01] is perpendicular to the shearing profile, it does not contribute to the velocity profile induced by the shearing. However, the flow of air does agitate the grains throughout the whole domain whereas our only source of agitation is the shearing velocity, the moving wall. Therfore in z-direction we put Dirichlet boundary conditions for the granular temperature. This is an attempt to simulate constant agitation of the grains. We will see that this is successful only on the boundary and is not able to model the agitation caused by the airflow in the experiment throughout the whole domain. As no absolute values of temperature are given in [BLS$^+$01] we set this agitation temperature as the mean value of the temperature between the moving and the resting wall from the measured results in [BLS$^+$01, Figure 3, pg. 5]. The top wall which corresponds to the outer cylinder in the experiment is a rough wall with no-slip boundary conditions. We set the velocity to be zero there.

The domain is discretized into $60 \times 40 \times 3$ volumes. Trying to match the parameters from [BLS$^+$01] as closely as possible, the internal friction angle is 30 degrees and the grains have a diameter of 0.75mm. The bare grain density is $2550 \frac{\text{kg}}{m^3}$. The shearing velocity u_{shear} in x-direction is $3.2 \frac{\text{mm}}{s}$.

3.2 Validation of the granular flow model

In this section we will try to validate the granular flow model with a few experiments. It is difficult to find real benchmark problems for granular flow

3.2. VALIDATION OF THE GRANULAR FLOW MODEL

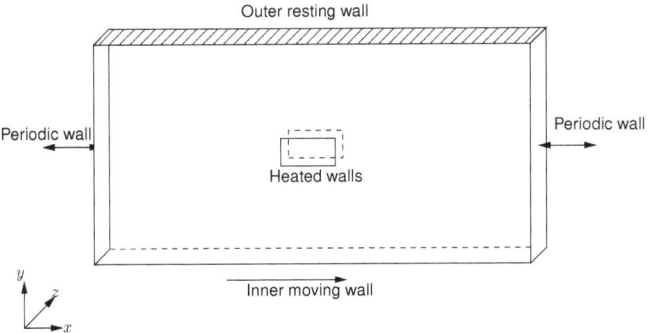

Figure 3.1: Setup of the numerical shearing experiment.

because granular material exists in very diverse flavors. As the modeling is not agreed upon, it is difficult to define a benchmark. We therefore validate using various experiments.

We begin with the results of the shear flow experiment which has been introduced in Section 3.1. We continue with the angle of repose which is an indicator for the internal friction angle, a material property. This property is studied in a Hele-Shaw-Cell which is a very thin cell between two plates simulating a 2D setup in a 3D material. Then we consider a phenomenon of granular flow which also relates to the internal friction angle. Granular material resists a force when put on an inclined plane. When increasing the angle of the plane the material suddenly starts to move. The dependency of the thickness of the resting layer on the inclination angle is studied and compared to an experiment. We continue with an experiment of dense and dilute flow around a cylinder and investigate the behavior of the flow. The last subsection considers the aspect of mixed derivatives in the granular flow model.

3.2.1 Shear flow

Let us present the results of the numerical shearing experiment and compare them to the experimental results from [BLS+01]. The velocity profile in Figure 3.2 matches with the one given in [BLS+01] which is the bottom left figure. There we match closest to the curve of solid triangles which is the experiment

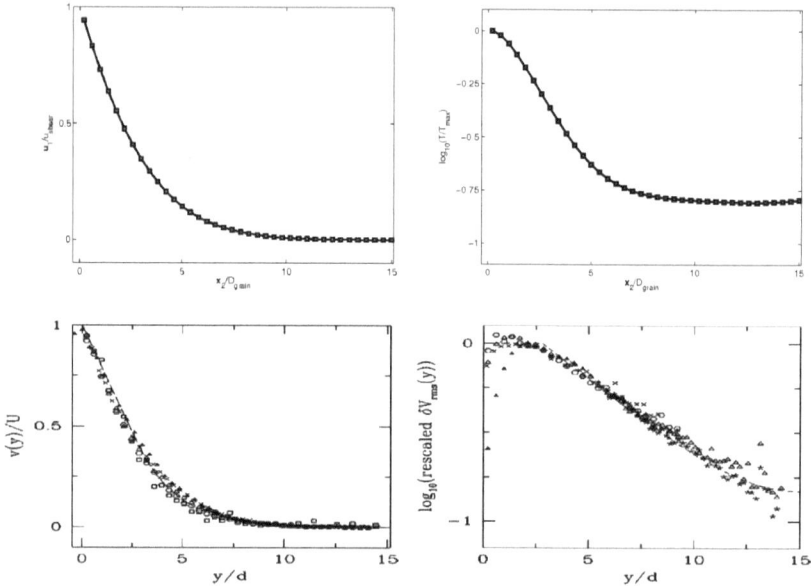

Figure 3.2: Top row: Velocity and temperature profiles of the numerical shearing experiment. The velocity u_1 is normalized by the shearing velocity u_{shear}. The temperature is divided by the maximum temperature and displayed on a logarithmic scale. Both values are displayed as a function of distance from the shearing wall normalized by the grain diameter D_{grain}. Bottom row: Figures [BLS+01, Figure 2, pg. 5] (left) and [BLS+01, Figure 3, pg. 5] (right).

3.2. VALIDATION OF THE GRANULAR FLOW MODEL   81

without airflow as we a computing it. The temperature profile however does not match so well with [BLS$^+$01, Figure 3, pg. 5]. We argue that this can be explained by the difference between the experiment and the numerical simulation and that the right graph in Figure 3.2 still makes sense for our setup.

We believe that the difference in the profile inbetween is caused solely by the influence of air blown through the whole domain in the experiment. Even though we apply an agitation source through Dirichlet boundary conditions for the granular temperature on the top and bottom walls, this is not sufficient to simulate the constant agitation caused by the airflow throughout the whole domain. And as there is no source of agitation other than the shearing wall, the fluctuation of the grains should decrease similar to the velocity profile.

This is why our temperature profile seems reasonable for our setup. Close to the wall, for a few layers of grains the temperature decreases very slowly as is the case in [BLS$^+$01]. Then it decreases quickly as the influence of the shearing decreases. When we reach the point where no velocity is induced by the shearing anymore, then the grains also reach their minimum agitation which is measured by the granular temperature. However, the amount of decrease in the temperature is matched perfectly again. At the outer, resting wall the measured temperature is, on the displayed logarithmic scale about -0.76 which is the same as in our simulation.

Let us further remark on the observation that we can find the same stability of the velocity profile as in [BLS$^+$01, Figure 2, pg. 5]. There it can be observed that the velocity profile remains almost unchanged for different shear velocities and different air flow configurations. We can observe in our simulations that the velocity profile also remains almost unchanged for temperature or periodic boundary conditions and for different velocity boundary conditions on the shearing wall.

From an analytical viewpoint, the temperature is discussed in [BLS$^+$01]. There, as usually for pure Couette flow, a solution with constant pressure is constructed. At constant pressure in [BLS$^+$01, Equation (20)] an analytical expression for the granular temperature is derived in [BLS$^+$01, Equation (25)] which agrees with the experimental results. This constant pressure solution

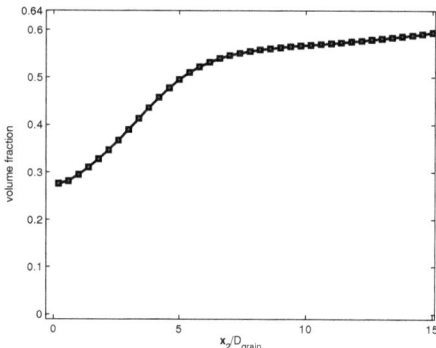

Figure 3.3: Volume fraction profile of the shearing experiment displayed as a function of distance from the shearing wall normalized by the grain diameter D_{grain}.

fixes the temperature at the moving and the resting walls. In our numerical experiment the wall temperature is prescribed at a value which is not necessarily consistent with the constant pressure solution because the absolute temperature values are not given in [BLS+01]. In this regard it is very interesting to observe the velocity profile does not seem to be influenced by our choice of wall temperature and hence not constant pressure.

The profile of the volume fraction is displayed in Figure 3.3. The authors of [BLS+01] do not give detailed graphs for volume fraction, but they do remark that the density close to the moving boundary is measured to be up to 40% below its limiting value. They further state that the volume fraction increases rapidly with distance from the sheared surface over several particle diameters. This corresponds to our observations in Figure 3.3.

3.2.2 The angle of repose

As discussed in Section 1.3.2 one of the most obviously observable properties that make granular media different from common fluids is the formation of piles. Even though the resulting angle of repose is not a material property, it is usually very close to the internal friction angle which quantifies the frictional interactions of the grains. We show in Appendix A.2 that for our hybrid model analog to [BLS+01] this internal friction angle is given by $\tan \Phi = \sqrt{\varepsilon_0 \eta_0}$.

So we may a priori choose a desired internal friction angle which the solution of the granular flow equations (1.18) should obey. Though the internal friction angle may differ by a few degrees from the measured angle of repose (which differs slightly through different experiments), the formula should still suggest a range of values to match the angle of repose found in our numerical experiments. We show that this is the case.

Setup of the numerical experiment

To measure the angle of repose we simulate a Hele-Shaw-Cell experiment. A Hele-Shaw-cell is a very thin rectangular domain which is completely closed except for an inflow at the top. We use a two-dimensional grid with 94×82 volumes with volume width of 1.25×10^{-2}m. The grid is uniform within the domain and is refined at the boundary to avoid that boundary effects obscure the results.

For initial conditions, we "fill" the domain with a volume fraction of sand of 1×10^{-4}. At the inflow we prescribe Dirichlet boundary conditions for the volume fraction (0.4) and the velocity in y direction ($-0.5\frac{m}{s}$). We should mention that the whole process is simulated in one run of solving the time-dependent equations described above. This includes the free falling of grains out of the inlet as well as the formation of the pile on the bottom.

Results

We run the simulation for three different internal friction angles of 51, 42 and 25 degrees by adjusting the value of ε_0. The measured angles of repose resulting from the simulations and displayed in Figure 3.4 come out as 48, 40 and 23 degrees. These values are within the commonly expected proximity to the internal friction angle.

3.2.3 Sliding down a rough inclined plane

Another aspect of granular flow is the transition from rest to flow depending on the force acting. An extensively studied experiment is the sliding of layers of granular media on an inclined plane. Depending on inclination angle, the

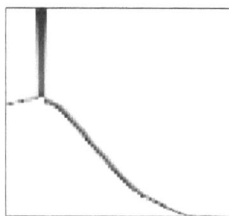 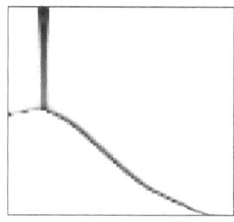 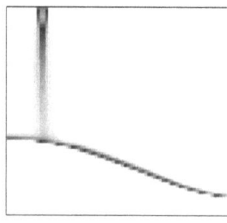

Figure 3.4: Inverse colour scale visualization of the volume fraction at the final stages of filling a Hele-Shaw-Cell with granular media at angles, from left to right, of 48, 40 and 23 degrees.

behavior of the granular media is quite different. Below the angle of internal friction, the grains either stay at rest or only a thin surface layer begins to slide. If the inclination is increased through and above the internal friction angle, the thickness of the layer of resting material decreases. In [BEL02], a kinetic model of granular flow is studied with regard to this aspect. A qualitative agreement to measurements is found there. Among many references, it cites [DD99] which gives experimental data which we want to compare our simulations with.

Setup of the numerical experiment

Similar to the Hele-Shaw-Cell experiment in Section 3.2.2 we study the flow in a thin cell, whose width is around 3 grain diameters of 1.75×10^{-4}m. The height of the initial resting bulk of grains is 7×10^{-3}m thick and rests on a 0.2m long plane. The resolution of the grid is $2 \times 10^{-3}m$ in x-direction and 5×10^{-4}m in y-direction. The initial conditions of resting sand are achieved by performing a filling simulation at zero angle. Then the plane is inclined which initiates the transition from rest to flow. The thickness of the resting layer is determined by declaring all grain layers with velocity less than $1 \times 10^{-3} \frac{m}{s}$ as resting. The inclination is actually achieved by a rotated gravity vector. The angle of repose is 25 degrees.

Results

Figure (3.5) collects the results of our simulations as functions of critical thickness against angle of inclination. The shape of the curve is in very good

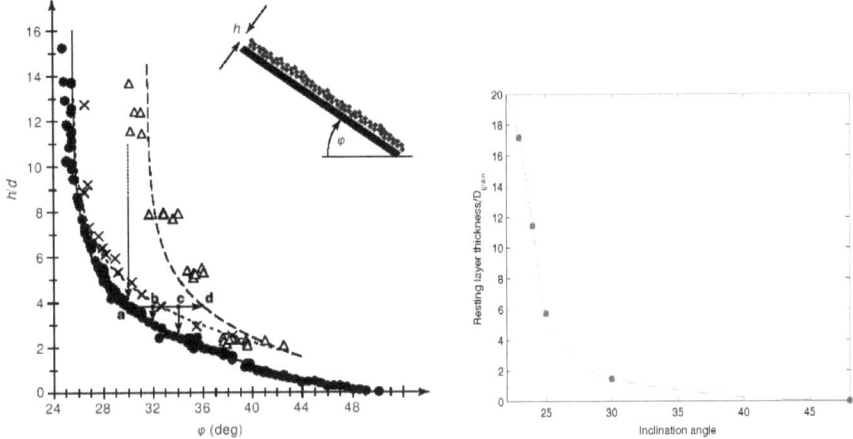

Figure 3.5: Plots of the thickness of the resting layer in number of grain layers against the inclination angle in degrees. Left: Figure 1 from [DD99] with experimental data from a sliding experiment. Right: Simulations have been run for the angles marked with blue rectangles, the graph is obtained by Matlab shape preserving interpolation.

agreement with the experimental findings in [DD99, Figure 1] even though we do not have quantitative agreement. This was expected as the reference does not give enough details on the granular material used for the experiments and we are not able to adjust all parameters to mimic the material used. However, it can be seen that in our numerical experiment the material immediately starts to move when the inclination angle is above the angle of repose of 25 degrees. There is also a strong point in favor of our hybrid model in contrast to the purely kinetic model. In agreement with experiments the thickness of the resting layer approaches zero for angles significantly above the internal friction angle, where in [BEL02] the thickness seems to become small but stays finite, see [BEL02, Figure 1]. The purely kinetic model proposed therein seems not to be able to mimic the granular material correctly in that regimes whereas ours does.

3.2.4 The stress tensor for granular flow

We want to follow up on the discussion of the form of the stress tensor σ (1.8) of the model for granular flow introduced in 1.3. There we have argued

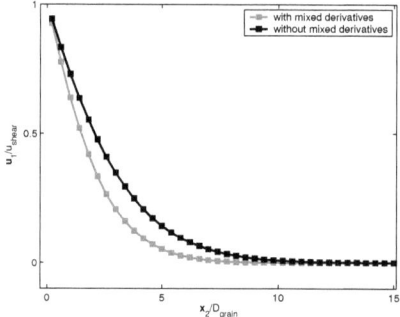 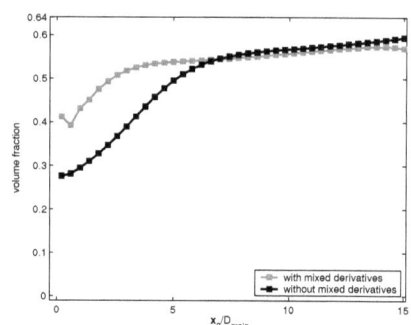

Figure 3.6: The plots show velocity (left) and density (right) profiles in the shear flow experiment from Section 3.2.1. The red line always shows the profile with mixed derivatives as in (1.1c). The black line shows the profiles with the stress tensor introduced in the granular flow model (1.18d).

that the specifics of granular media introduce a rotational viscosity and have ended up with a stress tensor that does not contain mixed derivatives like the general stress tensor (1.1c).

We want to strengthen our point by showing that the inclusion of mixed derivatives with κ as in (1.1c) leads to results which do not match experimental data. As a general remark, we have found in numerous numerical experiments that mixed derivatives cause a much too large diffusion of momentum. When piles are formed it shows that the granular media appears to be much too fluid-like.

We quantify this effect on the example of the shear flow experiment from Section 3.1. In the numerical experiments in Section 3.2.1 we have obtained velocity and density profiles which match the experimental profiles from [BLS+01]. We have carried out the same computation but including mixed derivatives using κ as in (1.1c). We compare the velocity and density profiles with and without mixed derivatives in Figure 3.6.

In the plots it can be observed that the results are quite different with mixed derivatives and more importantly, they do not agree with the experiment anymore. This together with other numerical experiments on the formation of heaps suggests that (1.18d) is the correct approach to the modeling of the stress tensor for granular materials.

3.3 Validation of the algorithm

We will investigate the numerical properties of the NPA introduced in Section 2.4. All the numerical experiments in the previous Section 3.2 are computed using NPA. This shows that the algorithm is generally able to compute solutions the granular flow model (1.18) which agree with experimental data. However, it is not clear if quantitative differences that occur at some points are due to the algorithm or are differences between the model and the real granular material used in the experiments.

We will first validate the NPA for the case of Newtonian flow in Section 3.3.1. It is certainly necessary that the nonlinear algorithm includes the case of Newtonian flow. Then we will look at the algorithm in the case of the full nonlinear granular flow model (1.18).

Analytical solutions to the model (1.18) exist only for the case of shear flow and there only in the high density limit. We have found agreement to our numerical results in that case in Section 3.2.1. Therefore, for validation we will compute the shear flow example on a very fine mesh in Section 3.3.2 and compare the solutions on successively coarsened grids to that. These studies also show that the periodic boundary conditions which are implemented as described in Section 2.5.1 work well for the NPA.

3.3.1 Newtonian flow

The NPA will be applied to the unsteady 2D benchmark from [TS96]. We run the unsteady test case "2D-2" from [TS96, Section 2.2 b)]. They consider the flow around a cylinder with circular cross section. A parabolic inflow profile is applied with a maximum velocity of $1.5 \frac{m}{s}$. The kinematic viscosity is defined as $\eta = 1 \times 10^{-3} \frac{m^2}{s}$ and the fluid density is $\rho = 1 \frac{kg}{m^3}$. This results in a Reynolds number of 100. At Reynolds numbers above 90 we usually expect some kind of unsteady flow behind the cylinder in the form of a Kármán vortex street. The benchmark geometry is displayed in Figure 3.7. A snapshot of the flow can be seen in Figure 3.8. When applied to an incompressible, Newtonian flow problem the NPA should reduce to a linear problem. As expected, the

CHAPTER 3. VALIDATION AND NUMERICAL SIMULATIONS

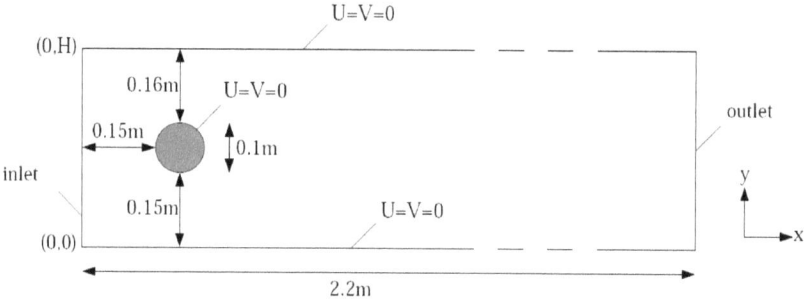

Figure 3.7: Geometry from [TS96].

Figure 3.8: Computations with NPA. Snapshots of velocity (top) and pressure (bottom) after a quasi-static state has been reached

number of Newton iterations to solve the pressure equation is 1. We want to compare our computation to the ones in [TS96]. One of their benchmark quantities is the difference in pressure before and after the cylinder. Because the flow is unsteady we have to take the time average of the pressure values after the flow has been established. In Table 3.1 we give the pressure drop averaged over 100 time steps. Our results are a bit outside the estimated interval for the "exact" pressure drop. This seems very understandable under the given circumstances. It is stated in [TS96, Section 5] that the most accurate solvers are using FV with contour adapted grids. We do not have

	Number of volumes	Pressure difference
NPA	40600	2.4463
[TS96]	$\approx 15000 - 300000$	Estimated "exact" interval: $2.46 - 2.50$ Actual results: $2.4587 - 2.5035$

Table 3.1: Comparison of NPA results for the benchmark taking into account only FV solvers.

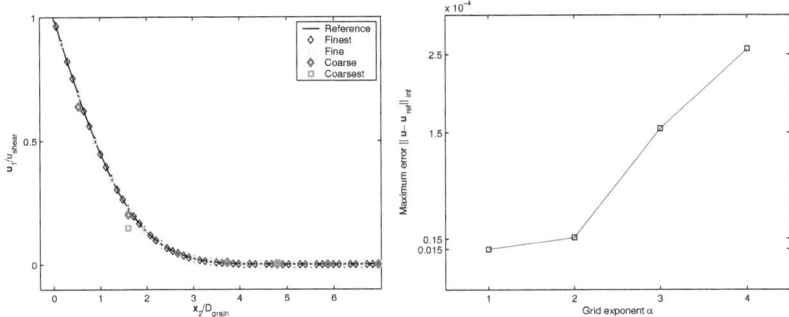

Figure 3.9: Left: We compare the velocity profiles of the shear flow experiment on different grids. Right: The maximum error of the solutions for the different grids is plotted agains the grid size. On the x-axis we plot the exponent α of the factor 3^α by which the reference grid is coarsened in each direction.

such an adaption. The cylinder in our case is approximated quite roughly by rectangular finite volumes. Furthermore, the results for the pressure drop differ very strongly across the different solution methods. Actually only the solver [TS96, 7a] is within the bounds for all grid resolutions. In the light of all factors taken into account in this benchmark our results seem very close.

3.3.2 Solutions for different grid resolutions

We investigate the solution of granular shear flow from Section 3.1 for successively refined grids using the NPA introduced in Section 2.4. Even for the shear flow, analytical solutions are only available for limit cases. It is even usual for Navier-Stokes Equations (NSE)-type systems that no complete analytical solutions are available for comparison to the numerical solutions. For simple partial differential equations (PDEs) this problem is resolved by applying the PDE system to a given solution and computing the right hand side. Then solutions for the system with this right hand side are computed and compared to the initially provided analytical solution.

For our very complicated system 1.18 this approach is not practicable. The equations are highly nonlinear. We cannot judge how stable the relations behave outside the physical realm of granular flow. So the task would be to find physically consistent profiles of density, velocity, pressure and granular temperature which satisfy their dependencies given in 1.18 and stay within

Grid	Number of volumes	$\|u - u_{ref}\|_{\inf}$	$\|u - u_{ref}\|_2 / \|u\|_2$
Reference	$648 \times 405 = 262440$	0	0
Finest	$216 \times 135 = 29160$	$1.540e - 6$	$5.578e - 4$
Fine	$72 \times 45 = 3240$	$1.684e - 5$	$7.145e - 3$
Coarse	$24 \times 15 = 360$	$1.553e - 4$	$8.244e - 2$
Coarsest	$8 \times 5 = 40$	$2.567e - 4$	$5.365e - 1$

Table 3.2: Errors of the computation on different grids with respect to the reference solution.

the existing limits, for example 0.64 for the density and conservation of mass. Even if we would find such profiles, a right hand side would appear in the momentum equation which acts essentially as a volume force. Therefore we approach the problem by computing a solution which we use as our reference on a very fine grid.

The reference solution is computed on a grid with the resolution of 648×405 volumes. The coarsest grid has 8×5 volumes and is successively refined by a factor of three in each direction which means that a volume of any grid holds 9 volumes of the next finer grid. We use the factor 3 because of our cell-centered grid. In this way we are able to compare solutions on different grids directly by restriction without interpolation because the centers of fine and coarse volumes coincide. In Figure 3.9 we show on the left how the solution approaches the reference solution through refinement of the grid.

The error values for infinity-norm and 2-norm together with the data of the different grids used is given in Table 3.2. Here we only compare the values of velocity, because we know from the experimental and analytical results in [BLS+01] that our computed reference velocity profile matches the data provided therein. We do not have any such data for the pressure.

The right picture in Figure 3.9 shows how the error is reduced by refining the grid. With each refinement of increasing the number of volumes by a factor of 9 we reduce the maximum error to the reference solution by roughly one order of magnitude. This is the case for all grids except from the coarsest to the coarse grid. There the error is reduced, but not by a whole order of magnitude.

3.4 Numerical investigations

In this section we will treat a few aspects of the algorithm development numerically. Most of the computations for the model (1.18) using the NPA 2.2 are carried out in Chapter 3. During the development of the NPA in Section 2.4 we have made a few statements about for example the conservation of mass and the compressibility. The numerical experiments here are meant to serve as supporting arguments for these statements and show certain properties which cannot be treated analytically because of the complexity of the model equations (1.18).

3.4.1 Compressibility regimes

At various stages in the development of the algorithm we have mentioned that the model (1.18) inhabits both compressible and almost incompressible regimes. This is clear from a physical viewpoint. A very dilute set of grains can be compressed very easily until it cannot be compressed at all anymore. To treat this analytically is very difficult because for the compressibility $\frac{\partial \rho}{\partial p}$ there is always the dependency of temperature, see (1.17). However, the temperature is part of the solution of the complete system (1.18) and therefore we can give the compressibility only for a certain solution.

So let us visualize the compressibility for a very simple numerical experiment. We choose the following setup. We hold a block of grains on the top half of a 3D box. At $t = 0$ we remove instantaneously the plane holding the grains. The computation is then started with gravity as the only acting force. Reality tells us the block of grains should fall to the bottom, compress there and rest at a certain packing fraction under the force of gravity. The different stages of the process are plotted in Figure 3.10

This shows that the compressibility varies strongly in both time and space between values of around 1×10^5 and 1×10^{-6}. Let us give a very rough sketch on Mach numbers. Of course, the speed of sound in solids or particle ensembles is a very difficult issue, but let us for this consideration assume we are dealing with a simple liquid. The speed of sound in liquids is defined as

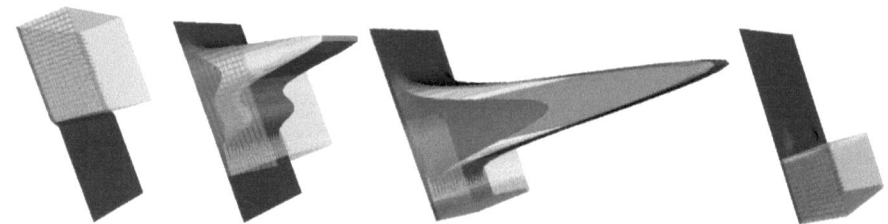

Figure 3.10: We visualize the process of the block of grains falling under the force of gravity. The transparent block is held on the top in the left figure, falls down in the two central figures and rest completely compressed on the bottom of the box in the right figure. The elevated plane shows the compressibility of the center layer in z-direction and varies between 1×10^{-6} (blue and flat) and 1×10^{5} (red, elevated) during the stages of the process. The grain density is visualized by the color of the transparent block where the green color means loosely compressed grains at volume fraction 0.4 and red means highly compressed grains marginally below a volume fraction of 0.64.

$\sqrt{\frac{\partial p}{\partial \rho}}$. So for our compressibility which is the inverse derivative $\frac{\partial \rho}{\partial p}$ the speed of sound ranges as low as around $3 \times 10^{-3} \frac{m}{s}$. For sand velocities of $1 \frac{m}{s}$ which are not uncommon we have Mach numbers of at least 100 which is highly compressible flow.

Our simulations suggests that our NPA 2.2 is able to handle these variations in a stable way. For linear fractional step methods (LFSMs) and specifically linear pressure correction algorithms (LPCAs) this is approached in various ways, for example in [BW98, vVW01] for highly compressible flows and [Chu03] for weakly compressible flows. It seems that in our case the use of a nonlinear pressure equation (NPE) makes these approaches unnecessary as we are not approximating the dependency of density on pressure.

3.4.2 Mass conservation

We have shown in the previous section that for the example of a falling block of sand, the compressibility varies strongly and that for granular flow we are in the regime of highly compressible flow with a large Mach number. With this in mind we want to investigate another property of the algorithm which we have claimed during its development. We have stated in Section 2.4.1 that the corrector of the NPA 2.2 ensures the conservation of mass even for highly compressible flow. Therefore we plot in Figure 3.11 the mass integral over the whole domain during the time-dependent solution process. The ex-

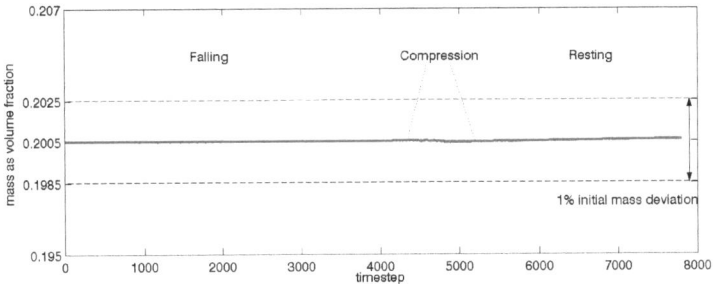

Figure 3.11: Mass integral of the falling grain block over time. The dashed lines plot an area of 1 percent of mass deviation around the actual mass curve. Note that at some point the sand hits the bottom of the box where it starts to compress.

periment suggests that the mass is conserved even during the period of high compression of the grains. Also, in the period after the compression, when the pile of grains rests without further movement is stable and the amount of grains that initially filled the box remains constant.

3.4.3 Bifurcating solutions

For the granular shear flow experiment from Section 3.2.1 we have found a property of our granular flow model (1.18) which is very interesting to mention, even though its thorough or possibly even rigorous investigation is outside the scope of this work. There seems to exist more than one solution in the granular flow model (1.18) and hence bifurcations. We are not the first to mention this. In [BEL02] the model from [BLS+01] which our model is largely based on, is applied to the situation of the flow on an inclined plane, compare also Section 3.2.3. The third paragraph in [BEL02, pg. 3] mentions the existence of multiple solutions above a certain volume fraction for certain parameters where only one is mentioned to be dynamically stable.

The curves in Figure 3.12 show that multiple solutions of density depending on temperature and pressure exist simultaneously at different points in space. The time-dependent simulation however starts with a bijective relation between density, pressure and temperature (1.17). Hence, there must exist a point in time, a bifurcation point, where a second solution branches off.

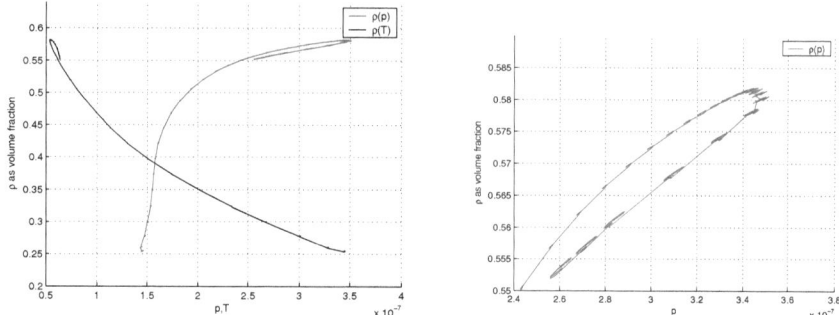

Figure 3.12: Left: Bifurcations in the solution for $\rho(p,T)$ in the shear flow experiment from Section 3.1. Right: Zoom in on the bifurcation point of the $\rho(p)$ graph.

In Section 2.1.1 we have provided arguments in favor of an implicit or semi-implicit algorithm. Bifurcating solutions are an aspect which might be handled better by an explicit algorithm. For an explicit algorithm it is more likely that the time marching stays on one of the branches once the bifurcation point is passed by the way the solution is computed. Basically an explicit algorithm starts with the solution of the previous time step and advances that solution with a very small time steps. In an implicit algorithm, the previous solution is only the starting point of an iterative procedure, in our case a Newton method in line 5 of Algorithm 2.2. Hence any local solution maybe obtained from this iterative procedure, especially as the solution branches have so little distance as in Figure 3.12.

3.5 Simulation of industrial processes

As the goal of work is not only the development of an algorithm and an implementation for validation purposes but also the development of a software, the code resulting code should be able to simulate granular flow on non-trivial domains. Also it should be easily extendable to solve new problems as they appear. In the Hele-Shaw-Cell experiment in Section 3.2.2 we have already shown that we are able to simulate a process of granular flow by filling the Hele-Shaw-Cell with a pile of grains.

We want to show here two further examples of the applicability of the code outside pure validation purposes. The first in Section 3.5.1 is the process of emptying a industry-size (2m high) 3D silo full of granular material. We carry out the complete simulation starting from the full silo until all granular material has flown out. In Section 3.5.2 we extend our granular solver for a very interesting application of solid mechanics. In some industrial applications it is necessary to compact granular material after it has been filled into a domain. We simulate this process by adding a volume force term to the momentum equation (1.18b) for certain volumes and hence applying a force on the top of the bulk of granular material. We observe an increase in volume fraction of grains throughout the whole domain and we find that a static pressure inside the bulk builds up which resists further compactification.

3.5.1 Emptying of silos

Simulation of granular media finds a wide range of applications in the field of handling of bulk goods. The vast majority of simulations in this area is carried out using DEM methods, treating each grain as a separate particle. The downfall of this method is the amount of grains that would need to be simulated for a full silo. For a silo of industrial size, a realistic estimate is 1×10^9 to 1×10^{12} particles which is out of the reach of current computation equipment by many orders of magnitude. Certainly there may be effects that can only be simulated by accounting for single particle interactions, but those have to be restrained to much smaller scales. With our method, we can simulate the complete process of emptying an industry-sized silo in days on standard non-parallel workstation hardware.

A basic qualitative phenomenon of silo flow is the distinction of core and mass flow depending on the steepness of the silo cone. For flat silos, so called core flow occurs where the grains flow towards the center of the silo and only in the center flow towards the outlet occurs. For steep silos, the grains at every point in the silo flow downwards, no inverse cone in the center is observable and mass flow occurs. In industrial silos mass flow is desired to avoid resting of goods in certain areas of the silo. It is known from experiments with the silo geometries that we use, that for usual internal friction

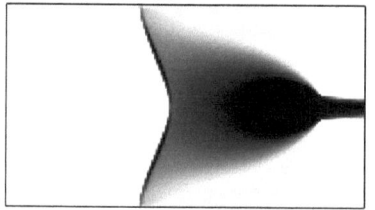 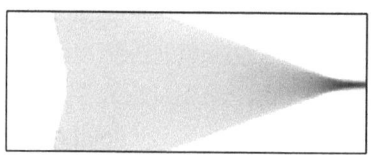

Figure 3.13: Plot of an intermediate stage in the process of emptying a silo with a 60 degree hopper (left) and a 20 degree hopper (right). We plot the magnitude of the mass flux where dark areas mean a large mass flux. The pictures are rotated for display convenience.

angles around 30 degrees a slope of 20 degrees will lead to mass flow, and at 60 degrees one should expect core flow. The slope is the angle between the hopper wall on the bottom of the silo and the y-axis.

We simulate the flow out of two different silos, one at 20 and one at 60 degrees. The shape of the silos can be seen in Figure 3.13. Both silos are discretized using 88×487 volumes. As the flow is mainly in y-direction, downwards out the silos, the domains are refined in that direction. Both silos have a radius of 0.35m and a height of 2m.

For numerical reasons, the domain does not end at the outflow as would be suggested by Figure 3.13. In fact, a channel with the radius of the silo is attached to the bottom of the silo which at its end contains the actual outflow boundary conditions. This has proven to be a much more stable outflow configuration. The top of the silo is also numerically treated as an outflow, it is an open boundary.

Another point which needs discussion are the initial conditions. We fill the whole silo with a volume fraction of grains of 0.5. This is below the maximum packing fraction which, because of the gravitational force, causes the material to compress before it actually flows out.

Our results show exactly the expected behavior in Figure 3.13. The difference between mass flow and core flow can clearly be seen. In addition we observe in the simulation that in the case of mass flow, the flux vectors are directed parallel to the direction of gravity at every point in the silo where in the case of core flow the grains on the top slide into the middle of the silo and only from there slide downward.

3.5.2 Compactification of granular media

One of the industrial applications for the compactification of granular material is the process of making sand cores for the casting industry, see [GKC96]. A sand core is created by forming a mold from a sand mixture either by shooting or filling sand in a form. In both cases the resulting sand might be compressed further by an externally applied force.

We present an approach to the simulation of compacting granular material using only hydrodynamic equations. We show that a locally applied force extends into the whole domain that is being compacted, and that distinct density and pressure distributions arise. Furthermore, we will show that the compacted state is stable, i.e. the flow is compacted from one equilibrium reached through pure gravity force to another equilibrium reached by applying the compactification force.

The flow of granular material is modeled by System (1.18). To apply the force we simulate a bar moving towards the sand filling which should compress the material. The movement of the bar of a certain density ρ_B is described by an ordinary differential equation (ODE). By χ_{Bg} we denote those volumes where the bar touches grains that are above a certain volume fraction c_{crit}. The bar moves with an initial velocity u_0 and is slowed down by a viscosity η_B. The equation describing this motion is:

$$\rho_B \partial_t \mathbf{u} = \rho_B \mathbf{g} - \eta_B \mathbf{u}_B - \sum \nabla p \chi_{Bg}. \qquad (3.1)$$

Equation (3.1) is a simple ODE which is solved by an implicit stepping algorithm. We write

$$\rho_B \frac{\mathbf{u}_B^{n+1} - \mathbf{u}_B^n}{\tau} = \rho_B \mathbf{g} - \eta_B \mathbf{u}_B^{n+1} - \sum \nabla p \chi_{Bg}, \qquad (3.2)$$

which gives

$$\mathbf{u}_B^{n+1} = \left(1 + \frac{\tau \eta_B}{\rho_B}\right)^{-1} (\mathbf{u}_B^{n+1} + \tau \mathbf{g} - \frac{\tau}{\rho_B} \sum \nabla p \chi_{Bg}). \qquad (3.3)$$

As mentioned before, the bar applies a force on the granular material. This is modeled by a right hand side of equation (1.18b). The force the bar exerts on the granular material is given by the acceleration of the bar multiplied with its mass. Hence the term

$$-\rho_B \left(\frac{\mathbf{u}_B^{n+1} - \mathbf{u}_B^n}{\tau} + \mathbf{g} \right) \tag{3.4}$$

is added to the right hand side of equation (1.18b).

The iteration between solutions for the granular system (1.18) and the bar movement (3.2) is schematically given in Algorithm 3.1. The last step

---
Algorithm 3.1: Grain compactification
---
1. Solve a complete time step for the system (1.18) with right hand side (3.4)
2. Compute the new bar velocity from (3.2) $\mathbf{u}_B^{n+1}$
3. Move the bar by $\mathbf{u}_B^{n+1} \cdot \tau$ to the new location $\mathbf{x}^{n+1}$
4. If the bar location is below the location of the volume face in moving direction, then set the bar onto the next layer of volumes
5. Set the grain velocity of all volumes that have grain concentration above c_{crit} and coincide with the bar to the bar velocity
---

in Algorithm 3.1 is carried out to ensure that all granular material moves at least with the velocity of the bar as there should be no grains where the bar has passed.

The density ρ_B of the bar is $6000 \frac{kg}{m^3}$, the viscosity η_B restraining the bar acceleration is chosen as $36000 \frac{kg}{m^3 s}$. To achieve non-accelerated movement of the bar (in absence of grain contact), we choose the initial velocity as $\mathbf{u}_0 = \mathbf{g} \frac{\rho_B}{\eta_B}$. Above a critical volume fraction we apply a force to the bar from the grains. The value of this critical volume fraction c_{crit} is 0.6.

The initial filling of the domain is achieved by simulating gravity driven compactification of grains until a steady state is reached. The initial density distribution is obtained by letting the granular material fall into the domain by the force of gravity.

Algorithm 3.1 is then applied to this initial configuration with the initial position of the moving bar being significantly above the grains. The bar starts to move, collides with the grain piles, is slowed down and compacts the grains. The final stage where the bar has come to a rest is shown in Figure 3.14

The compactification actually occurs, i.e. the grains rest at a higher volume fraction after compactification. We show this in Figure 3.15.

3.5. SIMULATION OF INDUSTRIAL PROCESSES

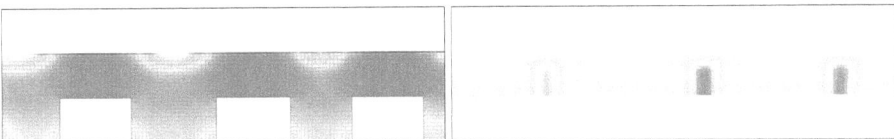

Figure 3.14: Pseudocolor plots of distributions of density and pressure after compactification. On the left figure it can be observed that the top of the bulk has been flattend by the force applied. On the right one can see the pressure buildup which resists further compactification.

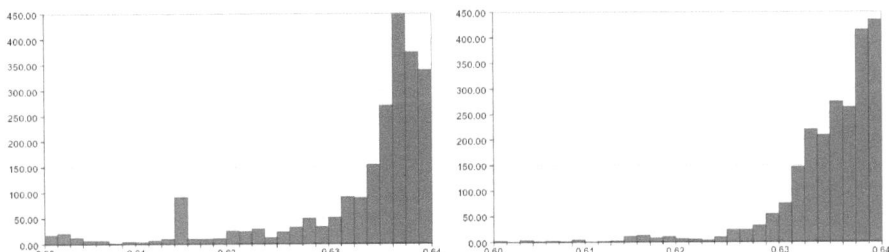

Figure 3.15: Histograms of density distribution before (left) and after (right) compactification. The horizontal axis holds the volume fraction and the vertical axis the number of volumes at this range of volume fraction. It can clearly be seen that the compactification process compacts the granular material in those volumes that had low volume fraction before.

Chapter 4

Software

A goal of this work is that the implementation of the algorithms from Chapter 2 goes in hand with the development of a software. By this we mean a code applicable beyond the scope of the thesis. There is a wealth of available computer programs for the simulation of fluid flow. In the following we call them computational fluid dynamics (CFD) codes. Our need to implement a completely new model and test different constitutive relations is the first motivation for developing our own code Complex Rheology Solvers (CoRheoS). The second motivation is the implementation of the novel nonlinear pressure algorithm (NPA) introduced in Section 2.4.5.

When solving partial differential equations (PDEs) for CFD, most codes start with implementing a discrete version of the system to be solved. Usually, the discretization is done beforehand and implemented in the code. In CoRheoS this is done by the code itself. We provide the software with the model in an abstract continuous form together with general rules how to discretize the parts involved. Any system of PDEs is a collection of possibly coupled equations of which each in turn is a sum of terms involving differential or algebraic expressions. We call these terms *parts*. For each such part a rule of discretization is given such that the program can do the work of actually creating a discrete system from the model.

We describe an approach to separate the CFD process into modules. We introduce a structure for programming which resembles the mathematical notation of equation systems and algorithms closely. We show that parallelization of the linear algebra and building of a matrix of the discretized problem can be developed independently of a specific problem. We make use of both the linear and nonlinear iterative solvers provided by PETSc version 3.0, see [BBG+01, BGMS97]. Regarding the notation, we depict diagrams in the UML format, see [SK98]. For code objects we use C++ notation. This includes denoting namespaces and class members by a double colon "::". For the complete C++ language definition, see [Str97]. Our approach to describe object oriented programming concepts using UML is quite common in computer science, see [Oes01].

The chapter is organized as follows. We first describe the modular approach that CoRheoS follows in Section 4.1. After at least mentioning all components we will omit the detailed description of the purely software-related ones such as visualization, input, output and framework design. We rather focus on modules that have a connection to the thesis. These are our approach to the discretization process in Section 4.2 and the parallelization of linear algebra components in Section 4.3. In the last Section 4.4 we give details on the multiphase aspect of CoRheoS.

4.1 Architecture and components

The design of CoRheoS is modular. We aim at developing a basis for rheology solvers rather than a single solver. Aside from a few specific demands, every flow solver needs basic components. Linear algebra components are nearly always involved in the solution process, even when the equations are nonlinear. Furthermore as soon as one is departing from academic computations the domain is always-non trivial. Therefore a grid needs to be generated from geometry input of some form. Undoubtly visualizion of the complete solution or aspects of it is important, therefore we need some way to output the results. All these components should only be coded once and then reused for many solvers. This is our aim for CoRheoS. Therefore we split up the CFD process into modular steps. They are given in schematic view in Figure 4.1.

The main separation is between what we call framework on the left side in Figure 4.1 and what we call implementation on the right. Following the well-accepted principle of programming to an interface, not an implementation, this separation appears both in the design of CoRheoS and in the actual code. The framework code is not touched by an implementation and all implementation code resides in a separate directory. Both connect only at compile time when the implementation code, which is a specialization of the implementation template, is compiled together with the framework code.

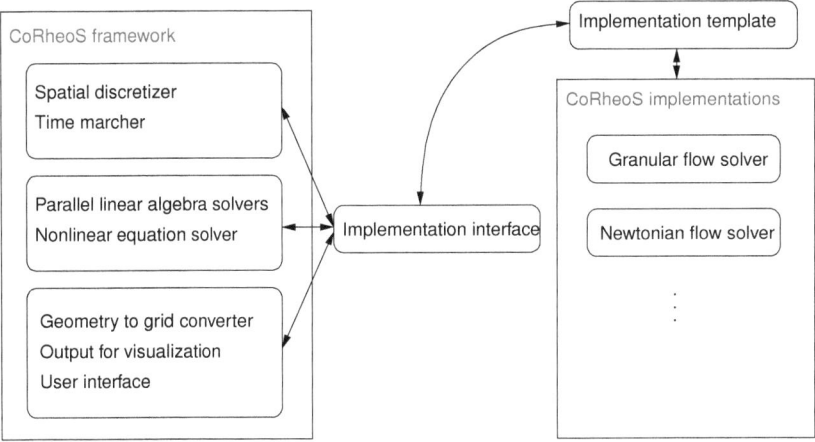

Figure 4.1: Components making up CoRheoS. A clear separation is sought between framework components common to all solvers on the left side and specific implementations on the right. Both are connected only by an interface.

4.1.1 The framework components

We mention shortly the components common to all implementations. By common we mean that all implementations may use the framework components as they are, but also may customize them to specific needs if necessary. This is done through so called hooks in the implementation template where an implementation may initialize a framework component with its own parameters.

Spatial discretizer

The spatial discretization component provides two things. First a structure to form an abstract description of a discrete system, together with discretization basics like interpolation and discrete spatial derivatives. Secondly it provides a facility to transform the abstract discrete system into a set of coefficients for each finite volume. To achieve this, it makes use of either standard builtin discretization routines or routines given in the specific flow solver implementation. These ideas are discussed in detail in Section 4.2.

Time Marcher

This component takes care of advancing the whole solution procedure in time, or, when solving steady problems it handles to pseudo time steps. It provides constant time stepping and the abstract class
corheos::TimeMarcher::TimeStepHandler
from which one may derive classes for different types of time stepping. Also it manages the storage of old and new values with respect to the current time step. We will not give further details on this component.

Parallel linear algebra solvers

The linear algebra component provides a basis for all solvers using matrices. It interacts with the spatial discretization component such that the coefficient sets after discretization are distributed among the number of parallel processes in a parallel matrix. Then this linear system is solved using parallel linear algebra solvers from PETSc. The solution is then collected from the various processes. Some of the ideas that this component is based on are discussed in detail in Section 4.3.

Nonlinear equation solvers

For solving nonlinear equation systems in CoRheoS we make use of the PETSc SNES library which provides Newton-like methods. It is not quite as straightforward to parallelize the solution process of a nonlinear system of equations as it is for the linear case. As of now CoRheoS provides only sequential solution methods for nonlinear systems. The idea however for parallelizing nonlinear solvers is clear. Two aspects have to be considered. The computation of the nonlinear function has to be split into multiple processes and the solution procedure of the linear system involving the Jacobian should use the parallelization approach that we will discuss in Section 4.3. The nonlinear aspect of CoRheoS is discussed in Section 4.3.3.

Input, output and user interface

These are the components on the bottom left of Figure 4.1. In CoRheoS, the generation of a Finite Volume (FV) grid from various geometry input formats is outsourced into the Geo2Grid library. This library makes of of the geometry data from CoRheoGrid written by Dr. D. Niedziela at the Fraunhofer ITWM which is able to transform arbitrary STL geometries to FV meshes. Within CoRheoS all its functionality can be accessed through the namespace geo2grid::. The visualization component makes use of VTK, an open-source visualization framework, see [SML03]. It provides the abstract classes corheos::VTKReader and corheos::VTKWriter as well as default implementations providing the reading and writing of CoRheoS results for VTK unstructured grids. The standard user interface is console-based but is itself only one of many possible implementations of the corheos::OutputFacility abstract base class. These components are of large interest for the software development aspect of CoRheoS but go beyond the scope of this thesis. Hence we will not give further details on them here.

4.1.2 The implementations

As mentioned before, implementations are separated from the framework through the implementation interface. This interface provides references to all CoRheoS framework components and binds the implementation to certain rules. Each implementation is a specialization of the implementation template as shown in Figure 4.1. Assuming some unique string as the implementation name, an implementation is declared by a configuration file called <implementationName>.config residing in the config folder of CoRheoS. A basic version of such a file is

```
CONFIG+=<implementationName>
<implementationName> {
        NAME=<implementationName>
        DEFINES += PHASE_COUNT=1
        DEFINES += PHASE_SINGLE=0
        IMPLEMENTATION_ROOT = <implementationName>
        IMPLEMENTATIONS += <implementationName>/implementation
        IMPLEMENTATIONS += <implementationName>/framework
}
```

Listing 4.1: A very basic configuration file for an implementation.

There are certainly more options for the file but for the scope of this work the above is sufficient. This assumes a folder structure

```
CustomImplementations/<implementationName>/
        declarations.cdm.xml
        framework/
        implementation/
```

Listing 4.2: Directory structure of an implementation

within the CoRheoS folder. An implementation following the structure in Figure 4.1 then consists of

1. The configuration file `<implementationName>.config` which mainly tells CoRheoS where to find the code of an implementation.

2. A set of function and variable declarations specific to that implementation in the form of a file called `declarations.cdm.xml`. This is essentially an XML file, see [xml98]. At compile time, the functions and variables declared here are converted to C++ code and merged with the framework code. This is done by the `cdm2c++` utility which is part of the CoRheoS framework.

3. A specialization of the implementation template in the `framework` subfolder. This is just a wrapper around the actual code mentioned below, but it implements all the functions declared in the implementation template.

4. The code of the implementation itself which is used by the specialization of the implementation template in the `implementation` subfolder.

4.1.3 Discussion of the modular approach

As mentioned above, we separate the different parts of the CFD process. The underlying design principles are not new and are well known in software design. In computer science, these things we have incorporated into CoRheoS are called *design patterns*. For a more detailed treatment of this topic we cite [GHJV95].

The idea in our design is to fulfill two goals simultaneously. The first is the separation of framework and implementation code. This is fully achieved by the above design. The second is to end up with a code that sacrifices as little performance as necessary. That is the more difficult part. One possible approach to separating framework and implementation would have been through inheritance. The framework would form base classes from which specific solver implementations inherit. However, the use of inheritance in components with many function calls can lead to poor performance.

This issue comes down to the ongoing discussion on performance in *inclusion polymorphism* versus *parametric polymorphism*. An example of the former is the use of inheritance, an example for the second is the use of templated classes or our own approach of inserting specialization functions directly into components. The performance loss that can be caused by realizing the design patterns in [GHJV95] by inclusion polymorphism is discussed in [DLGD01]. They favor *generic programming* which for us means parametric polymorphism or the use of templates or parameterized components as in our design of implementations above.

Even though from a software design point of view the discussion is quite straightforward, our highly object oriented approach is not without controversy in the mathematical community. We want to discuss the advantages and possible drawbacks of this approach. We start with what we think are strong points in favor of our approach.

- The reuse of code is maximized. If, for example the linear solver routines are general enough, they can be used by many kinds of solvers whether they are meant to solve different kinds of PDE systems or other types of problems.

- The cross dependencies of components on each other are minimized. To give a few examples, changing the discretization does not touch the linear algebra if that is not specifically desired. Adding or removing a term from the continuous system does not touch the discretization process. The parallelization does not touch anything but the linear algebra. Changing the constitutive relations does not touch the discretization.

- Development can be concentrated on the solver rather than wasting time on rewriting over and over again things like visualization, grid generation etc.

- All implementations benefit from improvements of framework components. At the same time, as the interface is assumed to be stable, none or little changes have to be done on the implementation to achieve this.

- Implementations have to follow certain rules because they specialize the implementation template. Therefore they share a common look. A developer who has written one solver is very likely to quickly be able to read the code for another solver, written by someone else.

- Development can be specialized and easily distributed across many developers. An expert in linear solvers can improve that part of the framework without having to know about PDEs. The developer of a PDE solver however will benefit from that expert work.

However there are drawbacks or things that might at first seem to be drawbacks especially when writing code for mathematical or physical applications. They shall not be omitted but discussed here.

- Layers of abstraction are necessary to realize the modularity of the software. These may have the effect of hiding connections in the code to inexperienced viewers. The code can not be read in a sequential manner. The code is distributed across many files and a deep hierarchy of directories. Functionality of a class is distributed within a hierarchy of inheritances. Object instances are managed by a central facility which requires advanced debugging methods in following the path of execution of the code.

- Programming errors introduced in the framework effect all implementations using it. Therefore it has to be carefully balanced that implementations always use the newest version of the framework but do not sacrifice functionality by that. The interface has to be kept stable and changes in the interface have to be communicated.

It is our strong opinion that the advantages outweigh the drawbacks by far and with careful software development they may not even be relevant. A highly efficient code which is only comprehendable by a few, or worse, only the author is useless. Also a code which is not extendable is useless. A compromise between efficiency and reusability must be established.

Let us stress one more point. Certainly, our arguments do not hold when solvers need to be developed for very special purposes. It is clear that one can easily write a solver for specific cases, for example an elliptic problem with constant coefficients, which is both faster and more memory efficient than any possible CoRheoS solver. But in that case the specific code can hardly be extended, let alone reused for a more complex purpose. In that way, CoRheoS can be seen as a framework to easily implement various solvers. Should a certain algorithm prove to be useful then writing a special purpose solver outside of CoRheoS may well result in a faster and more memory efficient program.

4.2 A generalized approach to discretization

It is our strong opinion that Finite Volume (FV) discretization of linear or linearized Navier-Stokes Equations (NSE)-type systems can be generalized on a code level. Our aim is to provide a structure in which one can code equations as they appear on paper in continuous form and hide the tedious coding of the actual discretization on a lower level. In CoRheoS we realize this goal in the following way. First we assume that the grid on which discretization is performed is made up of volumes of which each volume knows how to discretize different terms for itself. We explain this in detail in Section 4.2.1. In Section 4.2.2 we introduce a hierarchy of classes that resemble the structure of a continuous PDE system as closely as possible. By creating instances of the therein provided classes one can assemble a system of PDEs which already includes rules for discretizing each term. We then describe the process which creates a set of coefficients from those discretization rules. All the coefficients of all volumes then make up the final matrix.

As a guideline through explaining the process, let us introduce an example. We want to sketch the steps from a continuous system of PDEs to the coefficient matrix. Assume we want to solve the following system of cou-

CHAPTER 4. SOFTWARE

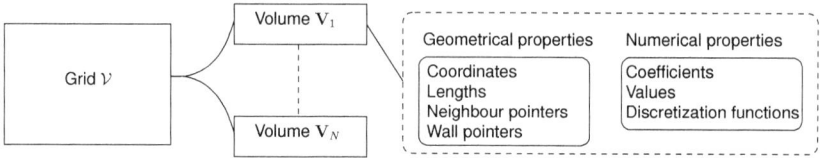

Figure 4.2: The datastructure of the CoRheoS grid. The grid holds a pointer to a list of volumes (CVs). These volumes hold properties like coordinates, but also values of unknowns on the grid and discretization functions.

pled PDEs. Using it purely for illustration purposes, let us assume a one-dimensional problem with a density ρ, a momentum m and $\alpha \in \mathbb{R}$.

$$\partial_t(\rho) + \operatorname{div}(m) = 0, \tag{4.1a}$$

$$\partial_t(m) + \operatorname{div}(\alpha m) = \text{rhs}. \tag{4.1b}$$

We do not worry about boundary conditions or constitutive relations.

4.2.1 Grid and volume data structure

In our code we try to resemble the grid structure introduced in Section 2.2. The grid in CoRheoS is made up of volumes which in the code we call *Control Volume (CV)*. Each volume has a certain number of faces (called *walls* in the code) depending on the dimension. The class representing a grid corheos::Grid holds a pointer to the list of volumes corheos::Grid::computatio * as displayed in Figure 4.2. Each volume holds geometric and numerical information. The geometric part consists of a pointer to a list of faces and a list of pointers to neighboring volumes as well as coordinates and lengths. The numerical information consists of an array of coefficients and discretization functions. The involved classes and their relationships are displayed in Figure 4.3. Let us explain in detail what we mean by numerical information.

In general, the grid relies on the functionality of the individual volumes. The idea is that when an operator is discretized on the grid, then the grid iterates over all its volumes and passes the work of discretization to each volume. This volume then knows how to discretize itself depending on whether it is in the interior, on the boundary or possesses some other property. Hence, discretization rules are stored in the volumes, as they may be volume depen-

4.2. A GENERALIZED APPROACH TO DISCRETIZATION 111

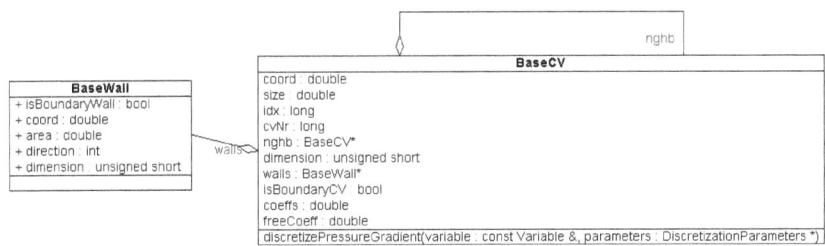

Figure 4.3: Simplified class diagram of a CoRheoS volume. In the actual code, there exist a few levels of inheritance. The most basic is the CV class declared in the geometry library from which finally the class `corheos::DomainCV2` and `corheos::BoundaryCV2` inherit. Displayed here is a stripped version of `corheos::BaseCV` which is common to both domain and boundary volumes. A sample discretization function is given in the CV class.

dent. When we say *coefficients* we mean the factors that appear in discretization in front of the central node of a volume and in front of the neighboring nodes. The discretization functions then fill the coefficients of every volume.

Coming back to our example, a discretization function for the term $\partial_t(\rho)$ is given in Listing 4.3. It handles the implicit Euler time discretization of the $\partial_t(\rho)$ term in our example system (4.1). It discretizes the integrated equations which is why the coefficients are multiplied by the volume. In a spatial discretization involving neighbors one has to then put the respective coefficients to the variables `coeffs[EAST]`, `coeffs[WEST]` and so on.

Every discretization function follows the general form of the above. Furthermore, this is all the work necessary to program a discretization. Everything else, from applying these rules to all the volumes of a grid to actually assembling a matrix from the values stored in the `coeffs` variable is done by the framework and will be described in the following section.

```
void
DomainCV2::discretizeTimeDerivative(const Variable & variable,
                    int component,
                    DiscretizationParameters * parameters) {
    coeffs[OWN]+=volume/timeStep[PHASE_SINGLE];
    freeCoeff+=-oldValues->
            retrieve(variable,component)*volume/timeStep[PHASE_SINGLE];
}
```

Listing 4.3: A sample discretization function for implicit Euler time discretization.

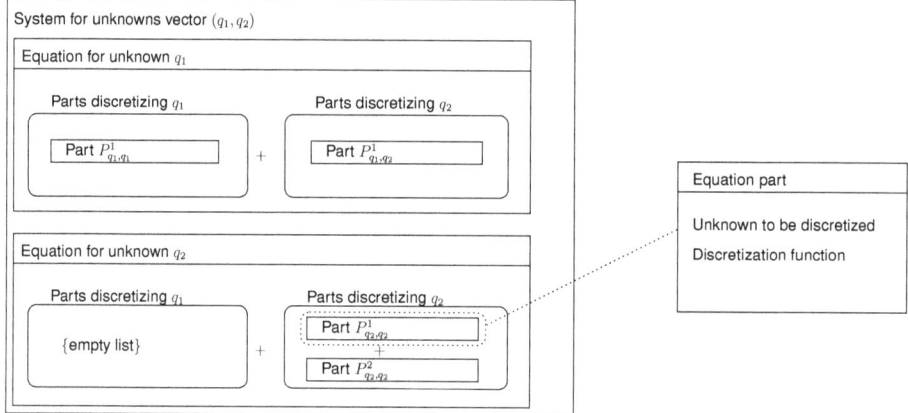

Figure 4.4: Scheme of discrete system in CoRheoS exemplified for our system (4.1). In general any system is a set of equations. Each equation is made up of sets of equation parts, one set for every unknown of the system. Each equation part provides a rule (function pointer) for discretizing that part.

4.2.2 The discretization process

We aim to automate the process to compute solutions of a continuous system of PDEs as much as possible. For this we split up the process into steps. Also we introduce a standard structure to describe systems. In CoRheoS, every model is considered as a system of equations. Each equation is made up of parts which in turn are made up of a variable and a rule of discretization of this part. Schematically, this is displayed in Figure 4.4. The figure 4.4 shows the correspondence of the discrete system in CoRheoS to the continuous sys-

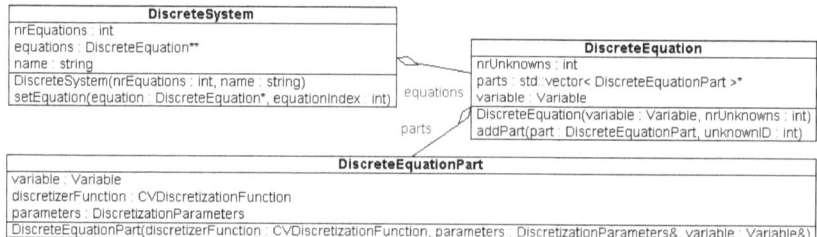

Figure 4.5: Class diagram of a discrete system. In CoRheoS all equations or systems of equations must be put in the form of corheos::DiscreteSystem. Each such instance must hold at least one corheos::DiscreteEquation. Each equation instance holds a list of instances of corheos::DiscreteEquationPart for each unknown. The discretization rules of all these parts then form the discretization and finally a coefficient matrix.

tem, for example (4.1). In that case q_1 is ρ, q_2 is m and the parts in Figure 4.4 refer to the terms in (4.1). In addition to the schematic view we provide the actual class diagram for the classes making up the discrete system structure in CoRheoS in Figure 4.5. We start with a system of PDEs in continuous form. The following steps then make up the discretization process.

- We describe the system in the form of Figure 4.4 which for example can have two equations and two unknowns q_1 and q_2.

- We apply Algorithm 4.1 which results in a matrix of the form

$$\left(\begin{array}{c|c} \sum P_{q_1,q_1} & \sum P_{q_1,q_2} \\ \hline \sum P_{q_2,q_1} & \sum P_{q_2,q_2} \end{array} \right) \quad (4.2)$$

where the summation of parts means that in every row, the coefficients resulting from every discretization function call are added.

Algorithm 4.1: Algorithm for discretizing a CoRheoS system: DISCRETIZER

1 Number of equations: N_{eq}
2 Number of unknowns: N_{un}
3 Number of volumes: N
4 Prepare matrix $\in \mathbb{R}^{NN_{eq} \times NN_{eq}}$
5 **for** $i = 1, \ldots N_{eq}$ **do**
6   **for** $j = 1, \ldots N_{un}$ **do**
7   Number of parts discretizing unknown q_j in equation i: N_{parts}
8   **for** $k = 1, \ldots N$ **do**
9   **for** $p = 1, \ldots, N_{parts}$ **do**
10   Call discretization function of part P_{q_i,q_j}^p in volume $\mathbf{V}_k$
11   Write coefficients from every volume to the respective matrix block $\sum P_{q_i,q_j}$ in (4.2)

We want to illustrate the process in actual CoRheoS code showing it for our example (4.1). To start, we need to formulate it in the discrete system classes. We assume that we have declared the variables `varDensity` and `varMomentum`, have available a set of default discretization parameters and provided the general rules to discretize time derivatives, a divergence and volume terms. Then the definition of our system in CoRheoS looks like the following.

```
DiscreteEquation eqnDensity(varDensity,2);
eqnDensity.addPart(new DiscreteEquationPart(discretizeTimeDerivative,parameters,varDensity),0);
eqnDensity.addPart(new DiscreteEquationPart(discretizeDivergence,parameters,varMomentum),1);

DiscreteEquation eqnMomentum(varMomentum,2);
eqnMomentum.addPart(new DiscreteEquationPart(discretizeTimeDerivative,parameters,varMomentum),1);
parameters.prefactor=alpha;
eqnMomentum.addPart(new DiscreteEquationPart(discretizeDivergence,parameters,varMomentum),1);
eqnMomentum.addPart(new DiscreteEquationPart(discretizeRhs,parameters,NULL),0);

DiscreteSystem system(2,"example");
system.setEquation(0,eqnDensity);
system.setEquation(1,eqnMomentum);

discreteSystems[system.name]=system;
```

Listing 4.4: Sample code for describing a system.

On a sidenote, the issues of boundary conditions and such are not relevant at this point. When the discretization functions of a volume are called, they know whether the calling volume is in the interior or on the boundary and then call the respective routines. In a next step, both the `corheos::Discretizer` and `corheos::Solver` need to be called on the system. The initialization of the linear system in the following line 1 is actually done in the `corheos::Initializ` component where the lines following that are put into the `corheos::TimeMarcher` component.

```
1 globalVars->registerLinearSystem(2                      ,"coupled");
2 DiscreteSystem system = algorithm->getDiscreteSystem("example");
3 discretizer->discretizeSystem(system,globalVars->getLinearSystem("coupled"));
4 solver->solveDiscreteSystem(system,globalVars->getLinearSystem("coupled"));
```

Listing 4.5: Code for solving a discrete system.

Using Algorithm 4.1, the `corheos::Discretizer` creates the coefficients for each volume and then stores the coefficients in the respective matrix blocks of (4.2). This matrix is solved using PETSc.

4.3 Parallel linear algebra

The discretization process described in Section 4.2.2 is executed on the first of many possible parallel processes. In line 3 of Listing 4.5, the function internally uses an instance of `corheos:ParLinSysHandler` to distribute the coefficients of each block onto the process in which the respective part of the parallel matrix resides. The parallelization makes use of PETSc which uses for internal operations MPI, see [MF97]. We use MPI within our C++ context for an efficient algorithm which reduces communication cost across processes.

4.3.1 MPI data structures

MPI is a standard describing a set of subroutines that can be used to exchange data between processes. A process is a running instance of a computer program. Plainly speaking, every process runs the same program with a unique process number (called *rank*) in its own area of memory. The MPI routines are then used to communicate data between the multiple processes. The standard is designed to be independent of programming languages and is therefore very basic in the datatypes it supports. However, it provides routines to mimic the high-level datatypes that we use in CoRheoS.

For the parallel linear algebra component we want to distribute blocks of sparse matrices to various processes. As there is always an overhead involved in communication between parallel processes we want to reduce the cost by sending a whole block at once. To achieve this, we have to define such a datatype that represents a matrix block in the code, retrieve its layout in the memory and then create the respective MPI datatype for communication.

We first define a class which holds one row of a sparse linear system, including the right hand side in Listing 4.6. In Listing 4.7 we show how to "trick" MPI into communicating instances of classes. The functionality of handling high level data structures such as classes is not built into MPI. Therefore we have to teach MPI how the class is laid out in the memory. Line 1 of Listing 4.7 announces that the memory of the new type will be divided into a field of integers, followed by a field of doubles and another field of doubles. In lines 6 to 11 we retrieve the memory locations of the integer and double arrays relative to the starting address of `linSysRows` from line 4. Line 15 then defines a new `MPI` datatype which corresponds to our class `LinSysRow` from Listing 4.6.

```
class LinSysRow {
        public:
        int indizes[GlobalVariables::nrMaxCoeffs];
        double values[GlobalVariables::nrMaxCoeffs];
        double freeCoeff;
        LinSysRow() {
                for (int i=0;i<GlobalVariables::nrMaxCoeffs;i++) {
                        indizes[i]=-1;
                        values[i]=0;
                }
        }
};
```

Listing 4.6: A sparse linear system row. The right hand side is stored in the variable `freeCoeff`.

```
1  MPI_Datatype oldtypes[3]={MPI_INT,MPI_DOUBLE,MPI_DOUBLE};
2  MPI_Aint offsets[3];
3
4  LinSysRow * linSysRows = new LinSysRow[2];
5
6  MPI_Address(linSysRows,&offsets[0]);
7  MPI_Address(linSysRows[0].values,&offsets[1]);
8  MPI_Address(&(linSysRows[0].freeCoeff),&offsets[2]);
9  offsets[1]-=offsets[0];
10 offsets[2]-=offsets[0];
11 offsets[0]=0;
12
13 int blockcounts[3]={nrNonzeroCoeffs,nrNonzeroCoeffs,1};
14 MPI_Type_struct(3,blockcounts,offsets,oldtypes,&MPI_linSysRow);
15 MPI_Type_commit(&MPI_linSysRow);
16
17 delete [] linSysRows;
```

Listing 4.7: Creating an MPI datatype for sparse matrix blocks.

Because we may send arrays of any `MPI` datatype, we are now able to communicate matrix blocks across processes. As we will show in Section 4.3.4 this results in so little overhead that for our problems we can call the method efficient and the overall memory and computation time consumption of this approach can be neglected.

4.3.2 Assembly of the matrix

The discretization process from Section 4.2.2 would result, in the serial case in a matrix containing as many blocks as there are equations times unknowns. It remains to be explained how line 11 of Algorithm 4.1 is actually done in the parallel case. There we have to distinguish two different block structures of the matrix. The first is given by the structure of the PDE system to be discretized. For our example above, System (4.1) we have a 2×2 block structure of the matrix because we have two unknowns and two equations which looks like (4.2).

But our matrix is also distributed among different processes and the number of processes not necessarily coincides with the number of equations. Therefore we need to distribute the coefficients into the parallel matrix by using our MPI datatype introduced in the previous section. Let us assume we have N_{eq} equations, N_{un} unknowns and we have N_{proc} number of processes. Then the left side of equation (4.3) shows the structure given by the discretization process and the right side shows how the matrix is distributed across the processes.

$$\begin{pmatrix} \sum P_{q_1,q_1} & \cdots & \sum P_{q_1,q_{N_{un}}} \\ \vdots & \ddots & \vdots \\ \sum P_{q_{N_{eq}},q_1} & \cdots & \sum P_{q_{N_{eq}},q_{N_{un}}} \end{pmatrix} \begin{pmatrix} \text{Rows on process 1} \\ \vdots \\ \vdots \\ \text{Rows on process } N_{proc} \end{pmatrix} \quad (4.3)$$

Certainly in most cases the number of unknowns does not match with the number of processes or at least that cannot be assumed. Therefore only parts of the coefficients of the blocks for the first unknowns may go to the first process or, if we are using a coupled system with many unknowns on few processes the overlap may be the other way around.

Using the MPI datatype defined in Listing 4.7 we proceed as follows. Assume the discretization is done for one block $\sum P_{q_i,q_j}$ reaching line 11 of Algorithm 4.1. This block is stored in the memory in the form of an array of the type linSysRow from Listing 4.6. We then need to determine what parts of that block goes to what part of the parallel matrix. This procedure in pseudocode is given in Listing 4.8, shortened to the relevant parts for the sake of a clear presentation.

```
void ParLinSysHandler::distributeCoefficients(int blockIdxEquation, int blockIdxUnknown) {

    offsetRows = nrCVs*blockIdxEquation;
    offsetColumns = nrCVs*blockIdxUnknown;
    GetMatrixPartOnRank(A, start, end);
    matrixBoundsStart[rank]=max(offsetRows, start);
    matrixBoundsEnd[rank]=min(end, offsetRows+nrCVs);

    if (rank is initialization rank)) {
        for (all other rankIdx in (ranks without initialization rank) do) {
            nrRows = matrixBoundsEnd[rankIdx]-matrixBoundsStart[rankIdx];
            LinSysRow * linSysRows = new LinSysRow[nrRows];
            rowStart = matrixBoundsStart[rankIdx];
            rowEnd = matrixBoundsEnd[rankIdx];

            for (int i=rowStart;i<rowEnd;i++) {
                cvIdx = i-offsetRows;
                domainCV=innerCVs[cvIdx];
                linSysRows[i-rowStart].freeCoeff=domainCV->freeCoeff;
                domainCV->getCoeffs(linSysRows[i-rowStart]);
            }
            MPI_Send(linSysRows, nrRows, MPI_linSysRow, rankIdx, ...);
        }
    }

    if (rank is not initialization rank) {
        nrRows = matrixBoundsEnd[rank]-matrixBoundsStart[rank];
        rowStart = matrixBoundsStart[rank];
        rowEnd = matrixBoundsEnd[rank];
        LinSysRow * linSysRows = new LinSysRow[nrRows];

        MPI_Recv(linSysRows, nrRows, MPI_linSysRow, globalVars->rank_init, rank, ...);
        for (int i=rowStart;i<rowEnd;i++) {
            rowIdx = i-rowStart;
            A->setValues(i, linSysRows[rowIdx])
            f->addValue(i, linSysRows[rowIdx].freeCoeff);

        }
    }

    if (rank is initialization rank) {
        rowStart = matrixBoundsStart[rank];
        rowEnd = matrixBoundsEnd[rank];

        for (int i=rowStart;i<rowEnd;i++) {
            cvIdx = i-offsetRows;
            domainCV=innerCVs[cvIdx];
            LinSysRow linSysRow;
            linSysRow.freeCoeff=domainCV->freeCoeff;
            domainCV->getCoeffs(linSysRow);
            A->setValues(i, linSysRow
            f->addValue(i, linSysRows.freeCoeff);
        }
    }
}
```

Listing 4.8: Pseudocode for distributing coefficients onto the parallel matrix.

In Listing 4.8, A is the matrix and f is the right hand side vector. In MPI terminology, each process is called *rank* and there is one master process which has the so called *initialization rank*. The discretization process described in the previous Section 4.2.2 is done on the initialization rank. Therefore the coefficients reside in the memory in this rank. A parallel matrix will be assembled from these coefficients.

The function in Listing 4.8 has four parts. The first part is the initialization of the matrix part on the current rank. Then it sends out the relevant rows to the other ranks in the second block. The third block of code receives these rows on the other ranks and fills the relevant parts of the matrix. The last block is there because parts of the matrix reside on the initialization rank. Even though it could be included in the first two, it would be inefficient to send memory from a process to itself. Therefore no MPI is involved in the last block.

This function is called in line 11 of Algorithm 4.1 for every list of parts belonging to one unknown in one equation. This function also serves well to illustrate how programming in MPI is done. Remember that the same code runs on all processes with the only difference being that the values of rank are different and the memory is separated between the processes. Therefore we check for the values of rank in the code above and know that different processes execute different parts of the above code.

4.3.3 The nonlinear case

The nonlinear case is treated separately, because we do not make use of a coefficient matrix in that case. We rather need to build a Jacobian and a function $\mathcal{N}$. The parallelization of nonlinear equation solvers is a bit more tricky than in the linear case. Firstly, the PETSc nonlinear library is much less matured than the linear components. Therefore, there is no such thing as a standard parallel nonlinear solver. The parallelization has to be done manually, is in many aspects mere programming work and hence goes beyond the scope of the thesis.

The ideas however are clear. As we have seen in Section 2.4.5, the basic

Newton method relies on three mechanisms. These are the evaluation of the nonlinear function $\mathcal{N}$ at a know vector q^n, the computation of a Jacobian matrix $\mathcal{J}$ at the know vector q^n and the efficient solution of a linear system of the form $\mathcal{J}(q^n)(q^{n+1} - q^n) = \mathcal{N}(q^n)$. The latter looks analog the linear case for each Newton iteration. However, the Jacobian may have to be computed at many points during the Newton method and therefore distribution of coefficients onto the processes may become an issue.

Therefore in the nonlinear case one has to develop a more sophisticated method of parallelization where the nonlinear function should be decomposed onto multiple processes and the overlapping components must be communicated. This is a challenging topic in itself and will be postponed to future work.

4.3.4 Discussion on the efficiency of the parallelization

As mentioned before, our parallelization is only concerned with the linear algebra component. This, in a way is a very basic and naive parallelization. In any computer program one has to deal with two bottlenecks, the memory and the computation time. Both can be eliminated with parallelization. Our approach does not eliminate the memory bottleneck, we can only hope to reduce the overall computation time. If a problem is too large for the memory of one CPU then more sophisticated methods have to be used, for example decomposition of the domain across processes.

The advantage of parallelizing only the linear algebra components is that there is no difference between programming a serial or a parallel solver. Discretization is carried out on the first process and the other processes are then only used for solving the resulting linear system in parallel. We have described in Section 4.3.2 how this is done.

This clearly limits the cases where we can actually expect a speedup of our solvers when using multiple processes. These are the cases where the solution of a linear system takes up most of the computation time. In that case, however, we are measuring in large parts the scaling ability of PETSc. Figure 4.6 shows the computation times for the pressure correction equation

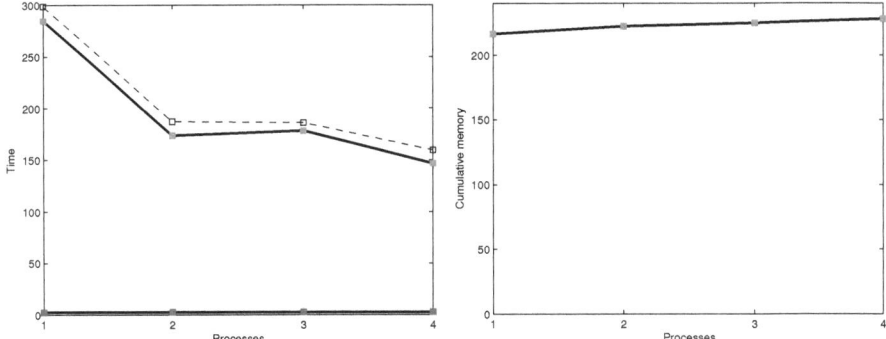

Figure 4.6: Left: The graph shows the CPU times (averaged over 5 identical runs) for 10 time steps of the benchmark problem from Section 3.3.1. The times are for solving the pressure correction equation (red markers) and for discretization (blue markers) The dashed line shows the overall computation time. The solution times include collection the results from all processes and the discretization times are for a complete run of Algorithm 4.1. Right: The graph shows the increase in memory caused by the parallelization. We see that the memory overhead is marginal.

summed over 10 time steps on 1 to 4 processes. Also in the graph are the times used for the discretization process. Taking into account that the cumulative time for the whole NSE time step (the dashed line in Figure 4.6) is only a few percent above the time needed for the solution of the pressure correction equation, we can see that the discretization process can be neglected with respect to computation time. Hence we do not create computational overhead by the distribution of coefficients onto parallel processes.

4.4 Multiphase

One of the advanced features and a major advantage of the CoRheoS framework is the possibility to combine existing singlephase solvers into a multiphase solver without changing any of the existing singlephase code and by adding only very little extra code. When we talk about multiphase, we mean the interacting or non-interacting flow of more than one materials in the same domain. This can be grains emerged in a fluid phase where for each material a different fluid model is used and the coupling happens through a term in the momentum equation. Or the flow of particles modeled as solid bodies that flow within a fluid phase and interact by volume forces on the particle side and an averaged force term on the fluid side. Solvers for flow of different materials can be made to interact very easily.

From now on when we use the term *phase* we mean an implementation of a flow solver as in Figure 4.1. This feature does not have a restriction on the number of phases but for the ease of presentation let us stick to two. As described in Section 4.1.2 an implementation consists of a few declarations, source code in specified folders and a configuration file.

Combining single phase solvers into a multiphase solver may best be explained using an example. Let us assume we have implemented an existing solver for a fluid phase called *phase1* and another for a fluid phase called *phase2*. So for both phases we have a configuration file very similar to the one in Listing 4.1. The steps to make a twophase solver from these not connected solvers, we create a directory twophase/framework and put in the necessary files. The configuration file `twophase.config` is given in Listing 4.9.

```
CONFIG+=twophase

twophase {
        NAME=twophase
        DEFINES += PHASE_COUNT=2

        IMPLEMENTATION_ROOT = twophase
        IMPLEMENTATIONS += twophase/framework
        IMPLEMENTATIONS += twophase/implementation

        DEFINES += PHASE_PHASE1=0
        IMPLEMENTATIONS += phase1/implementation

        DEFINES += PHASE_PHASE2=1
        IMPLEMENTATIONS += phase2/implementation
}
```

Listing 4.9: A very basic configuration file for combining single phases into a twophase code.

The work that remains towards a complete twophase code is the specialization of the implementation template. The easiest way to do this is to create a copy of the `CustomImplementations/implementationTemplate` into the `twophase/framework` folder. Then the function `timeStep()` of the `TimeMarcher` component has to call both the time step functions of the single phases and both phases are solved simultaneously. This is the most basic way to solve two phases but it is also a very trivial case. The phases are not yet coupled. Let us explain how this can be achieved.

4.4.1 Multiphase through coupling terms

The first very basic way to couple the two phases is to add terms to the systems of both phases and solve the systems one after another. This might be sufficient in the case of weak coupling of the phases. Explicit or implicit coupling terms are very easy to add to existing equations. We have seen in Listing 4.5 that a pointer to a system that we define in CoRheoS can be retrieved at any point through the class `corheos::Algorithm`. Listing 4.10 shows the necessary code to add an equation part to an already existing system. Assume that for both phases we have created discrete system as as in Listing 4.4 but have named them according to the phases *systemPhase1* and *systemPhase2*.

```
1 DiscreteSystem * systemPhase1=algorithm->discreteSystems["system_phase1"];
2 DiscreteEquation * discreteEquation=systemPhase1->equations[1];
3
4 DiscretizationParameters parameters = defaultDiscretizationParameters;
5 DiscreteEquationPart partCoupling(phase1DiscretizeCoupling, parameters, varMomentum);
6 discreteEquation->addPart(partCoupling);
```

Listing 4.10: Adding coupling terms to existing systems. The code shows adding a part to the system of phase1. The procedure is analog for the second phase.

The function given in the newly created part can be any discretization function as exemplified in Listing 4.3. Again no extra steps are necessary. After the terms have been added to the system, the code for each phase does not have to be changed, the discretization process will take into account the extra part automatically.

4.4.2 Multiphase through a coupled system

In case of a stronger coupling of the two phases, it might be necessary to not just solve the single phase systems with coupling terms one after another but to either rearrange the equations or couple both systems completely. As mentioned before, each system consists of one or more equations. For each phase these systems are already provided and we have access to the pointers to all parts of the structure. We want to build a new system which consists of the equations of both phases and can be solved in a coupled way. This can be done very conveniently by the code given in listing 4.11.

```
DiscreteSystem * systemPhase1=algorithm->discreteSystems["system_phase1"];
DiscreteSystem * systemPhase2=algorithm->discreteSystems["system_phase2"];
int nrEquations1 = systemPhase1->nrEquations;
int nrEquations2 = systemPhase2->nrEquations;
DiscreteSystem * twophaseSystem = new DiscreteSystem(nrEquations1+nrEquations2,"system_twophase");

globalVars->registerLinearSystem(nrEquations1+nrEquations2,"twophase");


for (int eqnIdx=0;eqnIdx<nrEquations1;eqnIdx++) {
        DiscreteEquation * equation = systemPhase1->equations[eqnIdx];
        twophaseSystem->setEquation(equation,eqnIdx);
}

for (int eqnIdx=0;eqnIdx<nrEquations2;eqnIdx++) {
        DiscreteEquation * equation = systemPhase2->equations[eqnIdx];
        twophaseSystem->setEquation(equation,nrEquations1+eqnIdx);
}
discreteSystems[twophaseSystem.name]=twophaseSystem;
```

Listing 4.11: Code to couple the systems of two phases.

In the previous case in Section 4.4.1 we were finished after modifying the system. In this case however there is some extra work to be done because none the singlephase codes knows of the newly created system. One has to add the lines to solve the newly created system to the time step function like in Listing 4.5 passing as parameters the created linear system and the discrete system.

The above examples should show that we have reduced to effort of creating multiphase code to a minimum. Most of the code written for each phase can be reused. If desired, changes in any of the singlephase codes can automatically be incorporated into the multiphase solver.

Chapter 5

Concluding remarks and outlook

The field of simulation of granular flow is large. We have seen that it touches a multitude of advanced topics. In modeling, these are for example the kinetic theory of granular gases and non-Newtonian fluid flow. Regarding the numerics we deal with complex Navier-Stokes Equations (NSE)-type systems and nonlinear discretization approaches. Therefore an in-depth and complete coverage of the simulation of granular flow within such a work is not possible.

We have presented a specific approach to the simulation of granular flow. We choose to model granular flow as a fluid in as many regimes as possible. We have introduced a hybrid model for dilute and dense granular flow which has a significantly simpler form than similar models introduced in the literature. Still, this modeling approach results in a very complex NSE-type system with all imaginable complications, varying viscosity, different regimes of very high and very low compressibility and nonlinearities in the algebraic relations.

For the solution of this System of partial differential equations (PDEs) and algebraic relations (1.18) we have developed a novel pressure based fractional step algorithm with a nonlinear pressure equation. The fact that we have not reduced the system and considered simplified versions but kept the full complexity of the system throughout this work is a contribution in itself.

To actually obtain approximate solutions of this model we have implemented the model as well as the nonlinear pressure algorithm (NPA) into a software framework which we developed in this work. This framework provides a generalized and automatic approach to discretization and automatic parallelization of the linear algebra components. Using this implementation we are able to compute solutions on arbitrary domains and visualize the results. The software however goes beyond the purpose of showing the applicability of the model and the algorithm. It is a general software framework for

CHAPTER 5. CONCLUDING REMARKS AND OUTLOOK

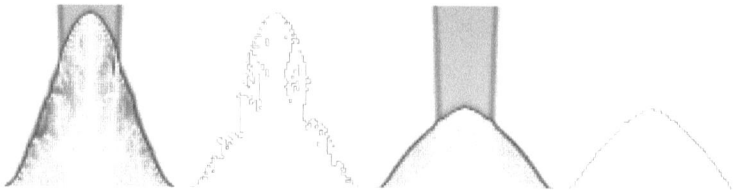

Figure 5.1: Instability in pile formation between $\beta = 1.75$ (first and second plot) compared to $\beta = 1$ (third and fourth plot). The first and third plots display the volume fraction from low (dark) to high (light). The contour plots (second and fourth) display the boundary between volumes where the yield relations for the dense regime (1.15) are active and those in the kinetic regime (1.11).

implementing solution methods for the flow of complex fluids. Our granular flow is just an example and a proof of concept for the software.

Using the implementation we have provided numerical experiments towards the validation of both the model and the algorithm. For the study of properties of the model and the algorithm we have carried out further numerical investigations. Finally we have applied the implementation to two industrial problems to further showcase the capabilities of the developed approach to the simulation of granular flow.

In the first chapter and in the introductions to the other chapters we have already given thorough overviews in the respective aspects of the simulation of granular flow. In the following we shall rather focus on insights and problems we found on the way.

Regarding the hydrodynamic modeling of granular flow, we can say that it is in some aspects a quite heuristic task. As of now there is no single correct constitutive model. That is not surprising because when we consider grains as hard spheres many obvious observations that we make when we look at, say, grains of sand, are left out. But not even all effects of few-particle systems with hard spheres can be incorporated without the equations becoming hopelessly complicated, see [BP03, pg. 34f]. Furthermore, minor specifics of the constitutive relations can have a major impact on the resulting solution. This is an important point in the modeling of granular flow and we want to emphasize it with an example.

During the development of the hybrid model we found a very interesting instability shown in Figure 5.1. In [BLS+01, Equation (15)] the viscosity (1.18h) is given with an exponent β on $g(\rho)$. The authors claim that a value $\beta > 1$ is necessary to obtain shear bands. When using our hybrid model this exponent is not necessary and it even causes the displayed instability. It is rather symptomatic for this field of modeling that one can get lost in the misery of constitutive relations, to cite [Kol00]. However, we believe that our model is as simple as possible with as few parameters as possible and, as we have shown, is still able to reproduce many effects of granular flow.

The modeling is followed by the task of computing solutions to the model. We have presented a pressure based fractional step method with a nonlinear pressure equation (NPE). We have explained why a linear method does not suffice in our case. After the decision for a NPE we followed the path laid out by linear fractional step methods (LFSMs). We ended up with two coupled equations for computing a new pressure and a new velocity in each time step. For reasons of stability we have used an upwind biased density in the mass conservation equation. The NPE was solved by a variant of the Newton method for systems of nonlinear equations. This resulted in a method which is able to solve the System (1.18). The method seems to be mass conserving and generally seems to be able to handle the huge range of compressibility quite well without any of the modifications which are necessary in the linear case, see Section 3.4.1. The rigorous analysis of these properties for the full system is not part of this work but we support the statements made in Section 3.4 at least numerically.

By using a NPE instead of a linear pressure correction equation (LPCE) we are able to obtain solutions of the very complicated granular flow model (1.18) but we introduce new problems. One is the question of when the Newton method converges for the NPE. It is well known that Newton methods only converge for starting values within a certain interval around the solution and have quadratic convergence only if all intermediate iterations stay within this interval. Modern Newton methods as introduced in Section 2.4.4 try to circumvent this restriction by very careful step size control but the issue of the initial value is still difficult.

This is especially the case for our way of solving the time-dependent problem. We use the value of the old pressure as an initial guess for the Newton method for the NPE. It seems that in most cases we are close enough to the solution for this approach to work. However, for a surely robust method we would need to investigate a sophisticated time adaption where we restart the computation for the current time step with a smaller τ if the iteration for the NPE has not converged.

This already brings us to the outlook of things that can and possibly should follow this work. The whole issue around the Newton method certainly gives rise to many possible directions of further research. We will not indulge into this further, as for this work we have been using a library for the actual solution of the NPE. For further research, it might well be necessary to investigate the optimal variant of the Newton method for our NPE from current publications and implement this from scratch. But there are some other points which can be of further interest. They are listed in the order of this work, first regarding the modeling, then the numerics and finally the implementation.

Static Coulomb friction: We have mentioned in the last subsection of Section 1.3.1 that the kinetic model has a property called dynamic Coulomb friction, investigated in [BLS+01]. This allows the formation of heaps close to an internal friction angle as we have shown for example in Section 3.2.2. We have extended the model for quasi-static situations and have kept the internal friction angle consistent as shown in Appendix A.2. Currently the angle of friction arises from the solution of the equations. In fully static situations, which are not part of this work, we have found problems in our numerical experiments. We suppose that for a situation of complete rest of the granular material, it is necessary to explicitely include a model of real friction of grains.

Partial slip boundary conditions: As we have mentioned in Section 1.3.3 the currently used boundary conditions are not completely realistic. It is necessary to model the interaction of the grains with the boundary in a better way than just assuming them to stick or slide freely on the boundary. We need to prescribe a friction for grain movement tangential to the boundary. This however makes it necessary to know exactly what direction is tangential to the boundary. Currently we approximate the boundary quite roughly by the cuboids.

Approximation of the boundary: There are different ways to find a better approximation of the boundary. Either at the point of grid generation for a certain domain one can form volumes at the boundary which represent the shape of the boundary. Another method is to use the same volume mesh but take into account the shape of the boundary during discretization.

Investigation of the NPE: With the NPE we have introduced a novel equation for computing an updated pressure value. In the linear case the elliptic equation, usually for a pressure correction is well studied. For the nonlinear case there exists the property of so called monotone operators which is a generalization of ellipticity.

Parallelization: In Section 4.3 we have described a parallelization approach for our generalized approach to discretization. This parallelizes only the linear algebra components. With this approach we do not remove the bottleneck of memory, we can only hope to speed up the computations. Some kind of domain decomposition should be considered. However, this may introduce new complications for the NPE. We would need to find a way to parallelize the solution of this nonlinear equation.

All the above topics should provide good starting points for further research. Aside from these points we have presented in this work an interdisciplinary and in itself closed approach to the numerical simulation of granular flow. It is closed in that it provides all the steps necessary to simulate granular flow. Aside from the aforementioned contributions in the separate parts of the thesis, this is the point where we see the major contribution of this work. The topic of simulation of granular flow has been treated in an interdisciplinary way. The three pillars of numerical simulation, namely the modeling, the numerical algorithms for obtaining approximate solutions of the model as well as the software implementation have been treated not separately but with respect to each other such that the final result is a way to simulate the behavior of dense and dilute granular flow in arbitrary domains.

"Who could ever calculate the path of a molecule? How do we know that the creations of worlds are not determined by falling grains of sand?"

VICTOR HUGO FROM LES MISERABLES

Chapter A

Appendix

A.1 Collaps in kinetic models

In Section 1.3.2 we described the shortcoming of the kinetic model in static situations without continuing energy input. To understand how the state of zero temperature and maximum packing fraction is reached we investigate Haff's homogeneous cooling for the temperature of the granular system described by Equations (1.7) and (1.11). Neglecting the gradients in the temperature and the velocities we obtain from (1.7)

$$\frac{\partial T}{\partial t} = -\varepsilon T. \tag{A.1}$$

For the final approach to a maximally packed state we can replace $\varepsilon(\rho, T)$ by ε^c in leading order with

$$\varepsilon^c := \varepsilon_0 \rho_c^2 \sqrt{T} g(\rho) \tag{A.2}$$

to obtain the equation

$$\frac{\partial T}{\partial t} = -\varepsilon_0 \rho_c^2 g T^{3/2}. \tag{A.3}$$

Equation (A.3) is solved by Haff's law (see [BP03, p. 53, Equation (5.5)])

$$T(\rho, t) = \frac{T_0(1 - (\rho/\rho_c))^2}{(1 - (\rho/\rho_c) + A\sqrt{T_0}(t - t_0))^2}, \tag{A.4}$$

with $A = \varepsilon_0 \rho_c^2$. Using this expression for T in the kinetic pressure p_K, compare Equation (1.11a), we see that the pressure vanishes with the density approaching the maximum packing state $\rho = \rho_c$. In addition the system becomes thermodynamically unstable. This can be seen by calculating $dp_K/d\rho$, which is for large times given by

$$\frac{dp_K}{d\rho} = \frac{1}{A^2(t - t_0)^2}\left(1 - 2\frac{\rho}{\rho_c}\right). \tag{A.5}$$

In the asymptotic long time limit for $\rho > \rho_c/2$, the compressibility would become negative for the pure homogeneous cooling case. Similar regions of negative compressibility have been found in other kinetic models of granular gases, see [KM02].

A.2 Dynamic Coulomb friction

That grains form piles is easy to observe and makes granular media different from common fluids. The angle at which the pile is formed, the angle of repose is not a material property, but it is usually very close to the internal friction angle. This internal angle quantifies the frictional interaction of the grains.

Bocquet et al. have shown in [BLS+01], that solid like behavior is a solution of the purely kinetic model which is System (1.18) with $\rho_{co} = \rho_c$ from (1.13) and (1.11) respectively. They derive that by solving with the kinetic expression for pressure and transport coefficients (1.11) for a shearing experiment at constant pressure, the shear stress σ_{xy} will be proportional to the normal stress component given by the pressure p independent of the shear rate

$$\sigma_{xy} = p \tan \Phi, \tag{A.6}$$

where the tangent of the internal friction coefficient is approximately given by

$$\tan \Phi = \sqrt{\varepsilon_0 \eta_0}. \tag{A.7}$$

We will now show, that we obtain the same relation with the choice of the hybrid model (1.14) and (1.15).

For an ideal shearing experiment the coordinate system can be chosen such that the changes of the velocity are perpendicular to the direction of the velocity. If we chose the velocity $\mathbf{u} = (u_x(y), 0, 0)$ the only non vanishing shear stress component is $\sigma_{xy} = \eta \frac{\partial u_x}{\partial y}$. Neglecting gradients in the granular temperature, Equation (1.7) gives the balance of viscous heating and dissipation as

$$\frac{2\eta}{3}\left(\frac{\partial u_x}{\partial y}\right)^2 - \varepsilon\rho T = 0. \tag{A.8}$$

Equations (1.8) and (1.9) give the ratio of shear stress and pressure

$$\frac{\sigma_{xy}}{p} = \frac{\eta}{p}\frac{\partial u_x}{\partial y}.$$

With Equation (A.8) this can be written as

$$\frac{\sigma_{xy}}{p} = \frac{\sqrt{\frac{3}{2}\eta\varepsilon\rho T}}{p}.$$

Using Equations (1.14) and (1.15) we obtain

$$\frac{\sigma_{xy}}{p} = \frac{\sqrt{\frac{3}{2}\eta_K\varepsilon_K\rho T}}{p_K}. \tag{A.9}$$

The angle of repose is relevant only for the high density limit, hence we replace ρ by ρ_c in Equation (A.9). Using the expressions from (1.11) we get

$$\frac{\sigma_{xy}}{p} = \sqrt{\frac{3}{2}\rho_c\eta_0\varepsilon_0}.$$

This shows that the yield pressure contribution does not affect the internal friction angle. Arguing that we are only interested in a rough approximation of the angle, we say that $\frac{3}{2}\rho_c$ is close enough to 1 to be neglected. Hence we arrive at the expressions (A.6) and (A.7).

The internal friction angle may differ by a few degrees from the measured angle of repose, which differs slightly through different experiments. Nevertheless, the formula should still suggest a range of values to match the angle of repose found in our numerical experiments.

Notation

Acronyms

CFD computational fluid dynamics

CoRheoS Complex Rheology Solvers A software framework for implementing flow solvers.

FV Finite Volume

LFSM linear fractional step method

LPCA linear pressure correction algorithm

LPCE linear pressure correction equation

MD molecular dynamics

NFSM nonlinear fractional step method

NPA nonlinear pressure algorithm An novel algorithm for pressure which involves a nonlinear pressure equation.

NPE nonlinear pressure equation

NSE Navier-Stokes Equations

ODE ordinary differential equation

PDE partial differential equation

SC Schur-Complement

Mathematical notation

<u>Vectors and vector valued functions</u> are typeset in bold font, regardless of whether they are continuous or discrete. Components of vectors are scalar, hence they are not bold as in
$$\mathbf{x} \text{ and } x_i.$$

NOTATION

We use the following differential operators in this work. The <u>divergence and gradient</u> are denoted by the differential operators div and grad. The indices i and j will denote rows and columns respectively. The divergence of a vector field $F : \mathbb{R}^d \to \mathbb{R}^d$ is defined as

$$\mathrm{div}\,(F) := \nabla \cdot F = \frac{\partial}{\partial x_j}(F_j).$$

For a <u>tensor valued function</u> $T : \mathbb{R}^d \to \mathbb{R}^{(d \times d)}$, $T = T_{ij}$, the divergence operator is defined as

$$\mathrm{div}\,(T) := \nabla \cdot T = \frac{\partial}{\partial x_j}(T_{ij}).$$

For a <u>scalar valued function</u> $S : \mathbb{R} \to \mathbb{R}^d$, the gradient is defined as

$$\mathrm{grad}\,(S) := \nabla S = \frac{\partial}{\partial x_j}(S).$$

Symbols:

ρ - Density, introduced in System (1.1)

$\mathbf{u}, \mathbf{v}$ - Velocities introduced in System (1.1)

p - Pressure introduced in System (1.1)

T - (Granular) temperature from Equation (1.7)

p_K, p_Y - Kinetic and yield pressure of the granular flow model System (1.18)

σ - Stress tensor introduced in System (1.1)

$\tilde{\sigma}$ - Stress tensor in the granular flow model System (1.18)

κ - Strain rate tensor introduced in System (1.1)

$\tilde{\kappa}$ - Strain rate tensor in the granular flow model System (1.18)

$\mathbf{q}$ - Granular heat flux from Equations (1.7) and (1.8)

λ, λ_K - Granular heat conductivity from Equation (1.15c)

η, η_K - Granular viscosity from Equation (1.15a)

$\varepsilon, \varepsilon_K$ - Granular energy loss rate from Equation (1.15b)

ρ_c - Limit density for random close packing of homogeneous spheres (1.12)

ρ_{co} - Crossover density between kinetic and yield regime, see Equation (1.13)

τ - Time step for time discretization (2.25)

NOTATION

π - Mapping between indizes of volumes, see Equation (2.3)

$\mathcal{V}, \mathring{\mathcal{V}}, \bar{\mathcal{V}}$ - Volumes of the grid, interior and boundary, see Equation (2.4)

$\mathcal{F}, \mathring{\mathcal{F}}, \bar{\mathcal{F}}$ - Faces of a volume, interior and boundary, see Equation (2.7)

$\mathbf{h}$ - Vector of lengths of finite volumes for space discretization, see Equation (2.5)

$\mathcal{DIV}$ - Discrete divergence, see Equation (2.13)

$\mathcal{G_F}$ - Discrete gradient by faces, see Equation (2.17)

$\propto$ - Denotes two proportional expressions

$< \cdot >$ - Averaging operator as in the definition of granular temperature (1.5)

$\mathbf{v} \otimes \mathbf{u}$ - Outer vector product $\mathbf{u}\mathbf{v}^T$

$\sigma : \kappa$ - Tensor contraction $\sigma_{ij}\kappa_{ij}$ in Einstein notation

Bibliography

[BBG+01] S. Balay, K. Buschelman, W. D. Gropp, D. Kaushik, M. G. Knepley, L. C. McInnes, B. F. Smith, and H. Zhang, PETSc Web page, 2001, http://www.mcs.anl.gov/petsc.

[BEL02] L. Bocquet, J. Errami, and T. C. Lubensky, Hydrodynamic model for a dynamical jammed-to-flowing transition in gravity driven granular media, Physical Review Letters **89** (2002), no. 18, 184301.

[BGMS97] S. Balay, W. D. Gropp, L. C. McInnes, and B. F. Smith, Modern software tools for scientific computing, ch. 8, Birkhauser Boston Inc., Cambridge, MA, USA, 1997.

[BH97] C. H. Bischof and P. D. Hovl, Using adifor and adic to provide a jacobian for the snes component of petsc, Technical Memorandum ANL/MCS-TM-233, Mathematics and Computer Science Division, Argonne National Laboratory, 1997.

[BLS+01] L. Bocquet, W. Losert, D. Schalk, T. C. Lubensky, and J. P. Gollub, Granular shear flow dynamics and forces: Experiment and continuum theory, Physical Review E **65** (2001), no. 1, 011307.

[BP03] N. V. Brilliantov and T. Pöschel, Kinetic theory of granular gases, Oxford Graduate Texts, Oxford University Press, Berlin, April 2003.

[BW98] H. Bijl and P. Wesseling, A unified method for computing incompressible and compressible flows in boundary-fitted coordinates, Journal of Computational Physics **141** (1998), no. 2, 153–173.

[Cam93] C. S. Campbell, Boundary interactions for two-dimensional granular flows. part 1. flat boundaries, asymmetric stresses and couple stresses, Journal of Fluid Mechanics Digital Archive (1993), no. 247, 111–136.

[Cho68]   A. J. Chorin, Numerical solution of navier-stokes equation, Mathematics of Computation **22** (1968), 745–760.

[Chu03]   A. Churbanov, A unified algorithm to predict compressible and incompressible flows, Lecture Notes in Computer Science, vol. 2542/2003, pp. 412–419, Springer, Berlin, Heidelberg, 2003.

[CL05]   M. Chuiko and A. Lapanik, Incompressible fluid flow computation in an arbitrary two-dimensional region on nonstaggered grid, Computational methods in applied mathematics **5** (2005), no. 3, 242–258.

[Dah59]   J. S. Dahler, Transport phenomena in a fluid composed of diatomic molecules, The Journal of Chemical Physics **30** (1959), no. 6, 1447–1475.

[Dar03]   S. Dartevelle, Numerical and granulometric approaches to geophysical granular flows, Ph.D. thesis, Michigan Technological University, July 2003.

[DD99]   A. Daerr and S. Douady, Two types of avalanche behaviour in granular media, Nature **399** (1999), 241–243.

[Deu04]   P. Deuflhard, Newton methods for nonlinear problems: Affine invariance and adaptive algorithms, Computational Mathematics, vol. 35, Springer, 2004.

[DLGD01]   A. Duret-Lutz, T. Géraud, and A. Demaille, Design patterns for generic programming in c++, COOTS'01: Proceedings of the 6th conference on USENIX Conference on Object-Oriented Technologies and Systems (Berkeley, CA, USA), USENIX Association, 2001, pp. 14–14.

[DMB+03]   G. D'Anna, P. Mayor, A. Barrat, V. Loreto, and F. Nori, Observing brownian motion in vibration-fluidized granular matter, Nature **424** (2003), 909.

[Duf01]   J. W. Dufty, Kinetic theory and hydrodynamics for rapid granular flow - a perspective, ArXiv Condensed Matter e-prints (2001).

[Dut88]   P. Dutt, Stable boundary conditions and difference schemes for navier-stokes equations, SIAM Journal Numerical Analysis **25** (1988), no. 2, 245–267.

[EG04]   A. Ern and J. L. Guermond, Theory and practice of finite elements, Applied Mathematical Sciences, vol. 159, Springer-Verlag, New York, 2004.

[EGH00]   R. Eymard, T. Gallouët, and R. Herbin, Handbook of numerical analysis, vol. 7, ch. Finite volume methods, pp. 713–1020, Elsevier Science, 2000.

[Fle91a] C. A. J. Fletcher, Computational techniques for fluid dynamics, 2nd ed., vol. 2, Springer-Verlag, Berlin, Heidelberg, New York, 1991.

[Fle91b] _____, Computational techniques for fluid dynamics, 2nd ed., vol. 1, Springer-Verlag, Berlin, Heidelberg, New York, 1991.

[FP96]   J. H. Ferziger and M. Perić, Computational methods for fluid dynamics, Springer, Berlin, Heidelberg, New York, 1996.

[GBHL06] L. Gastaldo, F. Babik, R. Herbin, and J.-C. Latche, An unconditionally stable pressure correction scheme for barotropic compressible navier-stokes equations, ECCOMAS CFD, 2006.

[GD99]   V. Garzó and J. W. Dufty, Dense fluid transport for inelastic hard spheres, Physical Review E **59** (1999), no. 5, 5895–5911.

[GHJV95] E. Gamma, R. Helm, R. Johnson, and J. Vlissides, Design patterns - elements of reusable object-oriented software, Addison-Wesley Publishing Co., Reading, Massachusetts, 1995.

[Gid94]   D. Gidaspow, Multiphase flow and fluidization, contimnuum and kinetic theory description, Boston, Academic Press, 1934, 1994.

[GKC96] A. Gardziella, A. Kwasniok, and L. Cobos, Recent studies comparing coremaking processes, Modern Casting **86** (1996), no. 3, 39–42.

[GMS06] J. L. Guermond, P. D. Minev, and Jie Shen, An overview of projection methods for incompressible flows, Computer Methods in Applied Mechanics and Engineering **195** (2006), no. 44-47, 6011–6045.

[GRS07] C. Grossmann, H.-J. Roos, and M. Stynes, Numerical treatment of partial differential equations, Universitext, Springer, Berlin, Heidelberg, 2007.

[GS78] B. Gustafsson and A. Sundström, Incompletely parabolic problems in fluid dynamics, SIAM Journal of Applied Mathematics **35** (1978), no. 2.

[GS98] P. M. Gresho and R. L. Sani, Incompressible flow and the finite element method. volume 1: Advection-diffusion and isothermal laminar flow, vol. 1, John Wiley and Sons, Inc, New York, NY (United States), December 1998.

[HA68] F. H. Harlow and A. A. Amsden, Numerical calculation of almost incompressible flow, Journal of Computational Physics **3** (1968), no. 1, 80–93.

[HA71] _____, A numerical fluid dynamics calculation method for all flow speeds, Journal of Computational Physics **8** (1971), no. 2, 197–213.

[HG97] J. S. Hesthaven and D. Gottlieb, A stable penalty method for the compressible navier-stokes equations. 1. open boundary conditions, SIAM Journal on Scientific Computing **18** (1997), no. 3, 658–685.

[Hir88] C. Hirsch, Numerical computation of internal and external flows, Wiley Series in Numerical Methods in Engineering, vol. 1, John

Wiley & Sons, Chichester, New York, Brisbane, Toronto, Singapore, 1988.

[HNS05] P. Hovland, B. Norris, and B. Smith, Making automatic differentiation truly automatic: coupling petsc with adic, Future Generation Computer Systems **21** (2005), no. 8, 1426–1438.

[Kad99] P. Kadanoff, Built upon sand: Theoretical ideas inspired by granular flows, Review of Modern Physics **71** (1999), no. 1, 435–444.

[KM85] J. Kim and P. Moin, Application of a fractional-step method to incompressible navier-stokes equations, Journal of Computational Physics **59** (1985), no. 2, 308–323.

[KM02] E. Khain and B. Meerson, Symmetry-breaking instability in a prototypical driven granular gas, Physical Review E **66** (2002), no. 2, 021306.

[Kol00] D. Kolymbas, The misery of constitutive modelling, Constitutive Modelling of Granular Materials, pp. 11–24, Springer, 2000.

[Lat06] A. Latz, Enthalpy equation: Temperature changes at constant pressure, ITWM Report, August 2006.

[LL78] L. D. Landau and E. M. Lifschitz, Hydrodynamic, Lehrbuch der theoretischen Physik, vol. VI, Akademie-Verlag, Berlin, 1978 (ngerman).

[MF97] MPI-Forum, Mpi-2: Extensions to the message-passing interface, Tech. report, University of Tennessee, Knoxville, Tennessee, 1997.

[MHN02] N. Mitarai, H. Hayakawa, and H. Nakanishi, Collisional granular flow as a micropolar fluid, Physical Review Letters **88** (2002), no. 17, 174301.

[Min01] P. D. Minev, A stabilized incremental projection scheme for the incompressible navier-stokes equations, International Journal for Numerical Methods in Fluids **36** (2001), 441–464.

[MP00] M.M. Massoudi and T. X. Phuoc, The effect of slip boundary condition on the flow of granular materials: a continuum approach, International Journal of Non-Linear Mechanics **35** (2000), 745–761.

[MPB03] B. Meerson, T. Pöschel, and Y. Bromberg, Close-packed floating clusters: Granular hydrodynamics beyond the freezing point?, Physical Review Letters **91** (2003), no. 2, 024301.

[NA07] M.-J. Ni and M. A. Abdou, A bridge between projection methods and simple type methods for incompressible navier-stokes equations, International Journal for Numerical Methods in Engineering **72** (2007), no. 12, 1490–1512.

[Oes01] B. Oestereich, Objektorientierte Softwareentwicklung - Analyse und Design mit der Unified Modeling Language, Oldenbourg, 2001.

[OS78] J. Oliger and A. Sundström, Theoretical and practical aspects of some initial boundary value problems in fluid dynamics, SIAM Journal of Applied Mathematics **35** (1978), no. 3, 419–446.

[PS72] S. V. Patankar and D. B. Spalding, A calculation procedure for heat, mass and momentum, International Journal of Heat and Mass Transfer **15** (1972), 1787–1806.

[QSS06] A. Quarteroni, R. Sacco, and F. Saleri, Numerical mathematics (texts in applied mathematics), Springer-Verlag New York, Inc., Secaucus, NJ, USA, 2006.

[RC83] C. M. Rhie and W. L. Chow, Numerical study of the turbulent flow past an airfoil with trailing edge separation, AIAA Journal **21** (1983), 1525–1532.

[Sav98] S. B. Savage, Analyses of slow high-concentration flows of granular materials, Journal of Fluid Mechanics **377** (1998), 1–26.

[SK98] M. Schader and A. Korthaus, The unified modeling language, Physica-Verlag, Heidelberg, July 1998.

[SK03] J. C. Sutherland and C. A. Kennedy, Improved boundary conditions for viscous, reacting, compressible flow, Journal of Computational Physics **191** (2003), 502–524.

[SML03] W. Schroeder, K. Martin, and B. Lorenson, The visualization toolkit, 3rd edition ed., Kitware, 2003.

[Str76] J. C. Strikwerda, Initial boundary value problems for incompletely parabolic systems, Ph.D. thesis, Stanford University, 1976.

[Str97] B. Stroustrup, The C++ programming language, 3rd edition ed., Addison Wesley Longman, Reading, Massachusetts, 1997.

[TS61] A. N. Thikonov and A. Samarskii, Homogeneous difference schemesa, Zhurnal Vychislitelnoi Matematiki i Matematicheskoi Fiziki **1** (1961), 5–63.

[TS96] S. Turek and M. Schäfer, Benchmark computations of laminar flow around cylinder, Notes on Numerical Fluid Mechanics, vol. 52, pp. 547–566, Vieweg, 1996.

[Tur99] S. Turek, Efficient solvers for incompressible flow problems, Lecture Notes in Computational Science and Engineering, vol. 6, Springer, Heidelberg, 1999.

[VPC96] P. N. Vabishchevich, A. N. Pavlov, and A. G. Churbanov, Numerical methods for unsteady incompressible flows using primitive variables and nonstaggered grids, Matematicheskoe modelirovanie **8** (1996), no. 7, 81–108.

[vVW01] D. R. van der Heul, C. Vuik, and P. Wesseling, Stability analysis of segregated solution methods for compressible flow, Applied Numerical Mathematics **38** (2001), 257–274.

[vVW03] _____, A conservative pressure-correction method for flow at all speeds, Computers & Fluids **32** (2003), no. 8, 1113–1132.

[Wes01] P. Wesseling, Principles of computational fluid dynamics, Springer Series in computational mathematics, vol. 29, Springer, Berlin, Heidelberg, 2001.

[xml98] Extensible Markup Language (XML) 1.0, 1998, http://www.w3.org/TR/1998/REC-xml-19980210.

VDM Verlagsservicegesellschaft mbH

Die VDM Verlagsservicegesellschaft sucht für wissenschaftliche Verlage abgeschlossene und herausragende

Dissertationen, Habilitationen, Diplomarbeiten, Master Theses, Magisterarbeiten usw.

für die kostenlose Publikation als Fachbuch.

Sie verfügen über eine Arbeit, die hohen inhaltlichen und formalen Ansprüchen genügt, und haben Interesse an einer honorarvergüteten Publikation?

Dann senden Sie bitte erste Informationen über sich und Ihre Arbeit per Email an *info@vdm-vsg.de*.

Sie erhalten kurzfristig unser Feedback!

VDM Verlagsservicegesellschaft mbH
Dudweiler Landstr. 99
D - 66123 Saarbrücken

Telefon +49 681 3720 174
Fax   +49 681 3720 1749

www.vdm-vsg.de

Die VDM Verlagsservicegesellschaft mbH vertritt

Printed by Books on Demand GmbH, Norderstedt / Germany